The Bee Friendly Garden

The Bee Friendly Garden

Easy ways to help the *bees* and make your *garden* grow

DOUG PURDIE

PREFACE

When I wrote my first book, *Backyard Bees*, I was focused on helping people who were keen to get into beekeeping. It was all about the goings-on inside a beehive and how to establish a honeybee hive and keep it healthy. Since then I've realised that there are lots of people who don't necessarily want to become beekeepers themselves but understand that our insect populations are under threat; they are aware that we face a very grey world if all our bugs are wiped out. When I do talks, these are the people who often ask, 'What can I do to help bees in my backyard?'

Increasingly, people want to know what they can do to provide forage for bees and how they can get involved in the movement to help our bees and other beneficial bugs. Hopefully, this book answers some of those questions and will help transform your backyard or balcony from an insect desert into a bug nirvana, as we save our bees one garden at a time.

Doug Purdie

Some particular bee-friendly plants include agastache (pages 2–3); borage (facing page); butterfly bush, lupins, asters, prairie coneflowers, artichokes and sedums (page 6); purple coneflowers (page 8); and lavender (page 9).

CONTENTS

THINK LIKE A BEE
11

The Birds and the Bees
25

The War on Weeds
45

THE WORLD OF BEES
59

Simple Changes Anybody Can Make
79

LET'S GET GARDENING
89

Edible Plants
101

NATIVE PLANTS
127

EXOTIC PLANTS
139

Beehives and Bee Hotels
157

**The Good, the Bad and the Ugly
—Other Garden Insects**
171

Conclusion—The Pollinator Highway
195

Appendix—Pest control recipes 200

Index 202

Acknowledgements 207

The world is a very different place when looked at through the eyes of a bee. This European honeybee is hard at work, foraging for nectar and pollen.

Chapter One

Worker Bee

THINK LIKE A BEE

It's a trick I use all the time when I'm talking to people about helping bees: I tell them to think like a bee. It's all too easy to view the world from the comfort of our two human eyes: to see what looks nice, what looks safe, what looks healthy. But it's a very different world looked at from the perspective of a bee with her five eyes and ultraviolet vision, on the constant hunt for nectar and pollen.

The first place where you need to think like a bee is at home. Pick up this book and step outside for a moment. Look around. How many flowers can you see? If you're like me and live in an urban centre, I bet you can't see many. Even in the suburbs you'll probably see more grass than flowers. This is the problem: we either plant grass or lay concrete with little regard for what our insect friends need. And the current trend for so-called 'architectural plants' isn't helping things at all.

In many parts of Australia there's also a building boom going on. What were once fallow fields of weeds and trees are being developed into unit blocks to house more and more people. Land is being released to developers and new suburban housing estates are being built, and Australia now holds the record for the biggest houses in the world—with potential garden areas sacrificed to make room for entertainment rooms and ensuites. Even in my super-dense neighbourhood, right at the city's doorstep, the sound of chainsaws and chippers is scarily common, and the honey yields from my home beehives are dropping dramatically. Our public garden spaces are under threat, with 100-year-old trees being removed to make way for more infrastructure. Governments promise that these trees will be replaced, but how can you compare a few new saplings with a tree that's been there for generations? All this is having an effect on nectar availability for all sorts of wildlife, including the all-important bees.

Last year the native Australian blue-banded bees (*Amegilla cingulata*) that had been visiting my

Bees lose out in the lawn-dominated, forage-poor landscapes of traditional suburban gardens.

house every year for the last 15 years disappeared. These are solitary bees that nest in dirt burrows, and I suspect that the soil they had been nesting in was dug up or concreted over—a sad end for a beautiful part of my local insect community.

The impact of the housing boom became very clear to me recently when I spent almost 12 months looking for a small factory unit to house our bee equipment and a small beekeeping supplies shop. Suitable commercial sites were incredibly hard to find, many having been bought so they could be demolished to build blocks of units—and furnish a more handsome profit for the landowner. I even heard of a complex of small industrial units that was only a few years old when it was demolished by developers to make way for lucrative new housing. My search eventually turned up a factory near the

Native blue-banded bees are disappearing from urban areas as their natural nesting sites—dirt and clay soils—are destroyed due to the pressures of growing cities.

airport, but even there a major development has begun, with a large number of native trees being removed to make way for a new multistorey hotel.

A lack of understanding of bees among tenants and property owners doesn't help. As an example, my business partner Vicky and I were once asked to attend some undeveloped industrial land that had become home to a whole aggregation of blue-banded bees. The caller was concerned the bees might sting him—mistaking them for European honeybees (*Apis mellifera*). That's disturbing, not only because it might have led to the destruction of harmless native insects, but because the bees were minding their own business, not hurting anyone, going about their pollinating. Luckily we convinced the owners to let the bees be.

THE PLIGHT OF CITY BEES

The problem for bees and other insects in urban areas can be broken down into three issues—loss, degradation and fragmentation.

The first one, habitat loss, is easy to recognise. In urban areas this means that lots of once open land or bushland is being cleared and turned into new commercial centres, housing developments and more infrastructure. Habitat loss is easy to recognise and understand—we can all see it around us—though sometimes we maybe don't recognise the extent of the impact it has on insects, birds and other wildlife that also claim the areas as home—if there is not enough to sustain them in the new surroundings they won't survive.

The second problem is harder to identify because we don't view the environment as bees or other insects do. But it's there: habitat degradation. An area that looks fine to us, with lots of things growing—maybe even a flower here and there—will, upon closer inspection (when you think like a bee) reveal that all is not good with the land.

Sometimes, too many introduced plants are the problem. And don't get me wrong—they're not all bad. We need a mix of flowering plants to support bees, and those introduced species can form the cornerstone of food supplies in the hard winter when there may be few native species around that are flowering. But the non-native species can also out-compete the natives for food, sunlight, space and water. Some native bees have evolved to need very specific things from their plant hosts—such as nectaries that are within reach if they have a short tongue, or a certain flower head design that gives them access to pollen grains—and introduced species might not provide these things.

To support all our beneficial insects, the mix of plants ideally contains more natives than introduced varieties.

Habitat can also be degraded through invisible means, like industrial pollutants and herbicides. All the chemicals in our environment affect it in some way, and some insects are very sensitive to so-called 'safe levels' of chemicals. This has been proven time and time again with banned chemicals like DDT (dichlorodiphenyltrichloroethane). As our reliance upon complex chemical cocktails increases and evolves, who knows what effect commonly used

So-called 'structural' and 'architectural' plants are popular in new developments, but provide no forage for bees.

Housing developments are causing habitat loss for our pollinators, but they don't need to.

insecticides will have on our bodies and the environment, or how long those effects will last.

The third issue, habitat fragmentation, is even harder for us to see.

It used to be that when a new highway was being built, the bulldozers would move in and just cut a swathe through the bush, lay down some asphalt and *voilà*—a road! At some point we realised that lots of animals needed to navigate this dangerous construction as well as humans. So for the past 30 years or so, in places like the north coast of New South Wales, the road builders have been installing underground tunnels and now aerial trapezes to allow ground and tree dwellers access to the other side of the road.

Now imagine a new housing development: row upon row of apartment buildings that produce deep, windy, dark canyons sometimes stretching out across whole suburbs. These developments are regulated to make sure that some perfunctory green space is planted or preserved between the buildings, which is great for humans and even

How do native bees find a way to survive in this new housing landscape or cross the distance to reach a park or nature reserve? They essentially become marooned.

the European honeybee (which is able to forage over many kilometres), but what about the native bees and other insects that inhabit much smaller areas and only travel a few hundred metres in their search for food? How do they find a way to survive in this new housing landscape or cross the distance to reach a park or nature reserve?

Think like a bee and you'll see that they essentially become marooned, and we know that this can't be good. Bees and other insects that try to adapt to these new conditions soon face all sorts of problems, the first being a lack of quality food, which leads to reduced hive populations that slowly become more susceptible to illness.

Green roofs have become a great way of using underutlised space to help cool a city in summer and provide some habitat as well. Instead of islands where insects become marooned, rooftops can be turned into archipelagos where bees can forage long distances, from rooftop to rooftop.

otherwise useful space that lies unused and only contributes by storing the sun's heat. Why don't we plant something?

Green roofs have become a great way of using unutilsed space to help cool a city in summer and provide some habitat as well. Instead of islands where insects become marooned, rooftops can be turned into archipelagos where bees can forage long distances, from rooftop to rooftop. Green roof installers are getting clever about the process, too—they're installing beehives to ensure they have pollinators on hand to keep these elevated ecosystems operating successfully.

THE BUTTERFLY EFFECT

Sometimes it's the plight of a creature perceived as beautiful that causes people to sit up and take notice. Unlike many insects that get a bad reputation because they sting or bite, the monarch butterfly (*Danaus plexippus*) is much admired as a magnificent and peaceful insect.

Every year, starting in September, huge populations of monarch butterflies migrate from southern Canada and the United States to Mexico, where they spend the winter, and then start the return trip in the spring, returning to their summer home in June. It's a natural phenomenon that people flock to see—the sight of millions of butterflies congregating in one place must be breathtaking and truly life-changing. (Monarch butterflies also migrate in Australian and New Zealand, but in nowhere near the numbers seen in North America.)

The life cycle of the monarch is as fascinating as the butterfly itself is beautiful. The butterflies that people see arriving in Mexico aren't the creatures that left Mexico the year before. Those original butterflies land in the United States, live in the local environment, mate, lay eggs, then die. Their offspring never migrate, spending their entire lives in the north, where they mate, lay eggs,

The other main problem is a lack of genetic diversity. Plenty of studies have shown how populations of any living thing that gets marooned or islanded from other members of its species become inbred due to a poor genetic mix. It even has a name—diffusion—and was famously described by Charles Darwin using finches as an example. The ultimate problem with diffusion is that the inbred populations are eventually genetically unable to breed with each other, and this can lead to extinction.

But the story doesn't need to end there. Fear not, we can reverse this.

WHERE THERE'S LIFE ...

As a rooftop beekeeper I get to see the city of Sydney from an elevated position and am constantly dismayed at how much grey concrete I can see up high. It's incredible to me that we have acres of flat, then die. And it's their progeny that migrates to Mexico—4000 kilometres away.

Recently, however, there have been fewer and fewer butterflies. Instead of a billion butterflies (which was the largest population ever recorded by the World Wide Fund for Nature in 1996–97), in 2013–14 the figures were the lowest ever, with only 33 million butterflies completing the journey.

WHY THE HUGE FALL IN NUMBERS?

The current theory about monarch butterfly decline attributes it to a lack of forage, primarily milkweed (*Asclepias syriaca*)—which is the only plant monarch caterpillars eat. This weed has been the brunt of a concerted effort by farmers to remove it using herbicides, along with all the other perceived 'weeds'. Another factor is habitat loss—caused by illegal logging in Mexico, wildfires and droughts. And if that weren't enough, the over-use of toxic chemicals, which pollute waterways and landscapes along the butterflies' migratory path, has contributed to these huge losses in numbers.

This same dramatic decline in insect numbers is taking place on a micro stage in our own neighbourhoods. When we look around our gardens, we've begun to notice how few butterflies and bees there are compared to years gone by. Their habitat is disappearing, and what remains is often doused with chemicals in our quest to eliminate feared weeds and pests.

Once upon a time, spring meant a familiar buzz of bees in the air; now there are few bees to be seen and backyard gardeners who try to grow zucchini or pumpkin discover, to their frustration, that the plants flower but no fruit appears. If they hand-pollinate they get fruit ... but that takes a lot of time. It's supposed to be done by bees—for free.

Without healthy populations of bees and other insects to pollinate plants, not only do wilderness environments come under threat, but also farmland. Greenpeace estimates that bees perform 80 per cent of pollination duties for all plants worldwide, and that 70 of the top 100 human food plants are pollinated by bees. Or to put it another way, 90 per cent of the world's plant nutrition is reliant on bees for pollination (with the remaining plants relying on other methods such as wind or other insects).

The good news in all this is that it's not too late for the monarch butterfly, and it's not too late for bees and other insects, too. In early 2014 the US, Canadian and Mexican Governments pledged to save the monarch butterfly, and President Barack Obama created a Pollinator Health Task Force to 'reverse pollinator losses and help restore populations to healthy levels', and to increase and improve pollinator habitat.

It's not just bees that are under threat. Monarch butterfly populations have dropped in huge numbers in recent years.

See the difference microbes make? The chemically treated field on the left is low in microbes, while the microbe-rich organic soil on the right is warmer and free of frost.

A RETURN TO THE OLD WAYS

All around the world people are waking up to the importance of bugs in keeping their gardens healthy. For decades, chemical fertilisers have been considered the best solution to growing healthy plants, but we're slowly coming full circle and gleaning that Mother Nature knows best ... who would have thought that the old-fashioned way was actually better, and that poo could be better at fertilising than phosphate?

Luckily, farmers are starting to consider how they treat their land and what they do with it. I love this photo (above) taken of adjacent fields owned by two different farmers, with the clear demarcation line at the fence. On the right, the organic farmer's land is rich in microbes, so is warmer and virtually frost-free, while the adjacent farmer's land has been treated with chemicals, and therefore has fewer microbes, and is covered in frost. This rethinking of so-called modern farming is an attempt to return our farmland to the lush bounty of the past and recognise that it's a complex ecosystem made up of all sorts of microbes, insects and plants— some of which might be just as important as the crop the farmer is growing and, if left intact, may even help crops to grow.

All of our insects have roles to play. Even the dreaded cockroach. Some, besides bees, are even pollinators. Before you lay down Armageddon on the bugs in your backyard, you need to consider what they contribute to our ecosystem. It's not just bees we need to save, but all pollinators and beneficial insects. So when you stand outside, imagine a backyard or balcony packed with flowering plants and, in the air, the chirp of crickets, the squawk of birds, the flutter of butterflies and, of course, the buzz of bees—just as it's meant to be. If you could understand 'Bee' they would be telling you they need help, and all your backyard creatures and critters are relying on you to do something about it.

Ladybirds play a vital role in gardens. As well as helping with pollination as they move between plants, many feed on aphids and scale insects that we consider pests.

Without the plants, pollinators will vanish. And vice versa—it's symbiotic. It's up to us to protect them both.

In the pollinator race the boldest wins. Daisy petals work like homing beacons guiding insects to the dinner table within.

Chapter Two

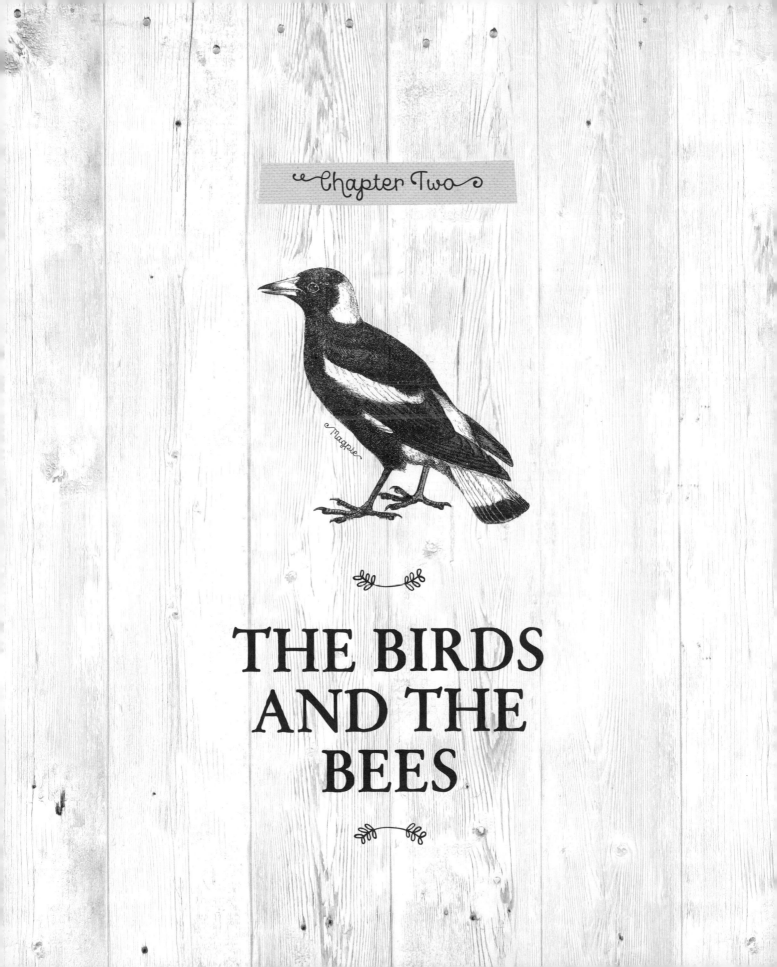

Magpie

THE BIRDS AND THE BEES

Over hundreds of millions of years, the evolution of plants was driven by the need to exchange genetic material with other plants of the same species, to enable reproduction and ensure genetic diversity.

Plants had a problem, however; they were evolving with their roots in the soil—they couldn't move. While some plants overcame this drawback and grew to pollinate themselves, others need pollen from another plant of the same species in order to produce fertile seeds and/or fruit.

Almonds, kiwi fruit, cucumbers, melons, apples, avocados, cherries (to name just a few of our food plants) all need pollinating to reproduce, and we need fertile seeds from our food plants to grow more of them. Even plants like lettuce and kale, whose leaves we eat; if they're not pollinated and don't produce seed, we can't keep planting them and reaping the rewards.

Let's look at the apple example: imagine an apple cut in half. Picture the seeds and segments. Each of those seeds is the result of pollination, and if one of the seeds isn't pollinated properly, then you'll have misshapen or smaller fruit, with seeds missing in the segments that have failed to develop.

If you've ever grown zucchini or pumpkins, then you've probably seen the result of bad pollination: a very promising fruit withers and falls off the vine,

This misshapen apple is the result of poor pollination.

much to your frustration, as you wonder what you did wrong. Well, it probably wasn't you; more likely it was a lack of pollinators in your vegetable garden. (You can test this theory by hand-pollinating a couple of pumpkin flowers—just make sure you exchange pollen from a number of flowers so you get good pollination.)

To understand pollination better, let's look at a flower and identify some major parts. About 90 per cent of flowers have both male and female parts. These flower are called 'complete' or 'perfect' flowers. The male part of the flower is called a stamen, which is made up of filaments (like stalks) with anthers at the top, containing pollen grains. The top of the female part (or pistil) is called the stigma. When pollen grains land on top of the stigma they grow a tube that travels down the style to the ovary. The pollen then fertilises female sex cells (contained in the ovules) to create an embryo, and the outer ovule and the ovary develop into the seed and fruit. (See diagram on page 29.)

Some plants have separate male and female flowers on the same plant. Pumpkin and zucchini are such plants—pollen is contained in the narrower male flowers (which have a stamen in the centre) and the female flowers have a more bulbous base which is the ovary.

Some flowering plants are able to fertilise themselves—with pollen being transferred from the male reproductive part to the female. This is known as self-pollination and these plants are known as 'self-fertile'. Lilies are an example of this. Other

Strawberry flowers are 'complete' flowers. At the centre is the pistil, containing the female reproductive parts. A ring of stamens (the male parts) surround it.

The sunflower family
Asteraceae

The sunflower family produces 'composite' flowers. They're actually made up of small clusters of flower heads packed together to look like a single flower. Daisies are part of the sunflower family.

Basic diagram of flower parts

Pistil
- Stigma
- Style
- Ovary

Stamen
- Anther
- Filament

Petal

Ovule

Sepal

Nectary

types of plants, such as mulberry trees, develop separate female and male plants, and both are required to achieve pollination.

For successful pollination, though, most plants need pollen from the flowers of another plant of the same species—this is called cross-pollination.

In an interesting turn, some plants that would otherwise be self-fertile have developed ways of preventing self-pollination, and these plants are referred to as 'self-incompatible'. These clever specimens seek pollen from other plants of the same species to prevent a genetic dead end. Some prevent self-pollination by only being able to receive pollen when they're not producing it. The avocado tree is an example of this. They have complete flowers, but the female and male parts open at different times to reduce the chances of self-pollination.

Successful cross-pollination requires a number of synchronised events. For starters, you need a source of pollen, called a polliniser. This polliniser needs to produce lots of good-quality, compatible pollen at the same time as the flowers that need this

pollen are flowering. If the timing is wrong and the pollen is early, or the flowers that need it are late, then no pollination can occur.

FRUIT TREES

All fruit trees need some form of pollination. Many fruit trees differ in their pollination needs, some being self-fertile and others cross-pollinating. For example, peaches are considered self-fertile because a crop can be produced without cross-pollination, though cross-pollination usually gives a better quality crop. Poor-quality pollination leads to low fruit yield and misshapen lower quality fruit. This can be a challenge for commercial growers.

Let's take another look at apples. They're a self-incompatible species, so they need pollen from a different type of apple tree to produce fruit. In a commercial orchard, many fruit tree varieties are

You need a mix of flowering plants with different shaped flower heads to make a good pollinator garden.

grafted clones rather than grown from seed—that is, while the rootstock is different, limbs from a specific variety are joined onto each. And if a single plant is used as the graft source, then each tree is genetically identical.

But this then leads to problems. If you allow these identical apple trees to provide the pollen for each other you'll end up with poor pollination and a poor crop. To counter this inbreeding, farmers often graft a different type of apple or a crab apple into every row to ensure genetic diversity and produce healthy fruit. (This is another reason to plant all sorts of plants in your garden: to assist with genetic diversity.)

BIOLOGY CLASS

So how does the pollen get from one plant to another when plants are stuck with their roots in the ground like wallflowers at a high school dance? There are two major ways for cross-pollination to occur.

The first is by physical means, known as 'abiotic' factors, such as being carried by the wind. Grasses (including food grains like wheat, corn and barley) rely on wind pollination. This group of plants includes about 8000 different species and tends to have small petalless flowers—which leaves pollen unprotected from the elements and easier to disperse. The downside for these types of plant is that wind pollination is a terribly imprecise and inefficient means of fertilisation. To ensure they get the job done, lots and lots of pollen needs to be flung around—which is fine if you're an ear of corn crying out for some cross-pollination, but not so great if you suffer from hay fever. (Incidentally, when the weather report says high pollen count it's usually referring to wind-borne grass pollen. Many people blame their hay fever on the trees they can see are flowering, but it's usually grass pollen that's to blame.)

The other method of cross-pollination requires 'biotic' factors—or living things—and means that some sort of insect, bird or animal does the job of transferring the pollen. About 75 per cent of all plants are biotic.

Just as wind-borne pollination is beyond the control of grasses, insects, birds and other animals don't deliberately help plants to pollinate. It's only when they accidentally rub against a flower stamen and pollen rubs off that they assist in the pollination process, but many flowering plants have evolved to manipulate a scenario so that this can happen.

EAU DE INSECTE

Plants have developed all sorts of tools or 'pollinator syndromes' to attract pollinators. These include scent, nectar and pollen treats, as well as alluring colours and petal shapes. These attractants may also change once the flower has been pollinated as a signal that the work has been done and there is no need to visit this flower again. Sometimes the smell of the flower dissipates, which makes it less attractive, or the nectar dries up so there is no reward for foraging bees. There is even discussion about some flowers having a static charge that's lost once a bee has visited, so that future bees will know that the flower has been recently visited by another bee and move on.

So let's have a look at these tools and see how they work to attract a plant's chosen pollinator.

Plants have nectar stores called nectaries (see diagram page 29). These produce the sticky sweet fluid that bees collect to make honey. Plants only produce nectar as an attractant to birds and insects, with wind-pollinated plants producing little or no nectar.

Along with nectar, some plants produce a strong fragrance also designed to attract pollinating insects. The scent carries on the wind and can

Flowers come in all shapes and sizes, some only suited to specific pollinators. From top left, clockwise: Artichoke, fuscia, dahlia and waterlily flowers.

Bees are the best pollinators by far, and this is for a number of reasons. The first being bee design—they have hairy little bodies that are very efficient at picking up pollen.

attract insects that can't see the flowers from far away; they recognise the scent as coming from a sweet flower. (This is also a good reason not to wear strong-smelling perfume or hair product around apiaries—the bees might confuse the smell with the fragrance of a flower.)

Other flowers have large shimmering petals as a 'come hither' sign to a passing insect, hinting at the rewards to be found. Colour is important as insects see light just as humans do, although bees see mainly in the ultraviolet spectrum so many colours we see are very different through their eyes.

It's all guesswork, but Karl von Frisch, the man who deciphered the bees' waggle dance (where bees direct others to good forage sites), also tested their ability to see colours. He found that bees could discern green, yellow, orange, blue, violet and purple. He did this by using coloured cards and food. He got the bees used to the idea that food could be found on a blue card, but not the other colours. When he removed the food, the bees still went to the blue card. He then tried this with green, yellow, orange, violet, purple and red. The only colour it did NOT work with was red.

Some plants are quite prescriptive about the pollinators they're trying to attract, down to the weight of the pollinator and length of tongue required to get at the pollen. Some plants (not many) are so specific that only one insect is able to pollinate them, placing both the insect and flower at great risk should either suffer decline. Fortunately, most plants rely on a mix of pollinators and most pollinators rely on a mix of flowers, so their success is almost guaranteed. That's a good thing if you like a variety of food in your diet.

It's also the reason I suggest later in this book that you should plant a broad variety of flowering plants—to attract the highest number of pollinators and also provide them with a varied diet, which is vitally important for their health.

BEES RULE AS POLLINATORS

Of all the external elements that make plant reproduction possible, bees are the most important. Bees are the best pollinators by far, and this is for a number of reasons. The first is bee design—they have hairy little bodies that are very efficient at picking up pollen. Pollen sticks to the hairs as bees move across a flower seeking out nectar and pollen treats. It's then randomly dropped off on the next flowers they visit.

The second is their methodical approach and complete devotion to the job of foraging. For the most part, bees will prefer to visit flowers from one species until that source has run out, at which point they then choose another. This is called constancy. You can see evidence of this if you look at a honeycomb pulled straight out of a beehive—its bands of colour correspond to the consecutive floral sources the bees have consumed, with each band in the row representing a different source discovered after the previous one was exhausted.

Constancy can lead to interesting consequences when the bees discover non-floral sources. This happened in New York in 2010 when a bright-red honey was suddenly being produced in hives around Brooklyn, creating (I have to say it) quite a buzz. The phenomenon was eventually traced to Dell's Maraschino Cherries Company in Brooklyn, a maraschino cherry factory that was, coincidentally, experiencing a problem with bees. The red honey incident was solved (and resolved), by installing bee-proof barriers in the cherry factory, and the bees went back to producing a more conventionally coloured honey.

I've had personal experience with this myself. I've taken phone calls from bakers in Sydney (who glaze their cakes with honey) complaining about bees visiting their shops and stealing the glaze from the cakes, much to the concern of the shoppers. You can only imagine the conversation back in the hive about the virtual paradise the forager bees have found.

Purple is an especially strong colour in the ultraviolet colour range. Salvia produces abundant nectar for bees.

Crepe myrtle flowers contain a pollen bonus for bees—they produce a false pollen just for the bees, to help digestion, and a second type which they use for fertilisation.

The bands of colour in this bar of honeycomb show constancy at work.

Of course, because of constancy the bees will keep coming until the source is removed, so the bees remembered the cake shop and kept coming back until the baker removed the cakes from the display.

I've also experienced constancy at home. My bees know that I often have a car that smells like honey from carrying hives and honey jars around, and they seem to appear the moment I arrive at the garage. Because of the risk of spreading disease from hive to hive, I don't let them get the honey, but nonetheless they always turn up to have a look and try their luck.

The third reason bees make such spot-on pollinators is that they visit flowers specifically to obtain pollen or nectar, unlike a lot of other pollinators that visit for a multitude of reasons. Some visitors, like cockroaches or beetles, are only interested in chewing the petals (not the pollen or nectar at all) and only by chance become pollinators.

Bees may visit hundreds of flowers in a single trip, making them very efficient at covering a large range and achieving high rates of pollination. They're looking for nectar or pollen, and clever flower design ensures they must push against the stamen or stigma to get what they're looking for, meaning there is a higher chance of pollination by a bee than, say, a butterfly, bird or caterpillar.

BUZZ POLLINATION

Some plants are so single-minded about maximising the pollen transfer that they require a special sort of insect action to release the pollen. This is called buzz pollination or sonication. About eight per cent of the world's plants require buzz pollination, including many food crops, such as tomatoes, potatoes, eggplants, blueberries and cranberries. These plants have the pollen locked up in a narrow tube, only releasing it when the tube is vibrated. Bees do this by vibrating their wing muscles at around 400 Hz while gripping the flower anther with their mouthparts, and they're very good at it.

Some of these plants can be partially wind-pollinated, but sonication ensures proper pollination and better fruit yield or seed production. In greenhouses in Australia, where bees do not have ready access, a sonic wand is often used as a human substitute. It's a bit like a torch with a probe that vibrates at the right frequency to release the pollen, but it's not as effective as bees.

A SHORT HISTORY OF BEEKEEPING AND BEE DISEASES

Humans have maintained bee colonies for thousands of years, with evidence that the Egyptians practised migratory beekeeping, floating beehives up and down the Nile River on rafts—we assume for honey production. But it wasn't until somewhere between the 1700 and 1800s that humans worked out how pollination worked and the part bees played in it. The first country to attempt commercial pollination using managed bees was New Zealand in 1885, using bumblebees (*Bombus* spp.) imported from the United Kingdom to pollinate red clover (*Trifolium pratense*)—the bumblebees' tongues were more suited to that crop than the imported European honeybee that was already well established there.

Keeping bees for pollination is now quite a lucrative money maker in many parts of the world, with many beehives trucked all over the place to provide pollination services for farmers. We don't see a lot of this in Australia because we have healthy feral honeybee populations, the biggest exception being almond pollination, which is a growing part of Australian agriculture that relies heavily on beekeepers for pollination. In the USA, since bee numbers have started falling due to disease, pollination has become big business, with a reported $200 per session paid for each beehive. When you consider some beekeepers have thousands of hives, the income generated can be huge.

Here in Australia we have a relatively healthy bee population that's not suffering the bee diseases that the rest of the world's beekeeping populations now face—namely, varroa (*Varroa destructor*) and colony collapse disorder (or CCD for short). Both have caused a major decline in honeybee numbers around the world and Australia is the only continent in the world where they haven't infiltrated.

Varroa is a mite about the size of a pinhead that infests beehives and lives on the heamolymph (or blood) of the bee. In the process of feeding it weakens the bees and, a bit like a mosquito, spreads a virus and other diseases to the colony. It's fatal in most cases and causes huge beehive losses when it first enters a country.

Varroa was first discovered in the early 1900s as a parasite on the Asian honeybee (*Apis cerana*), and spread around the world after European honeybees became infected in the 1940s.

CCD is something that's largely unexplained, but can be best described as a beehive losing most of its bees almost overnight, leaving behind a population that's unable to sustain itself and eventually dies. The reasons for CCD are unknown, but most probably are a group of things rather than one thing—that group being made up of loss of quality food sources, overuse of insecticides and bad management practices by beekeepers.

Another communicable disease related to bad hive management is American foul brood (AFB). This is a spore-based bacterium that's spread from hive to hive, usually by bees stealing the honey from a weakened, affected hive and taking it back to their own. This disease, unlike varroa and CCD, has hit Australian shores and is present in every state. The infection figures, thus far, are low, but Australian beekeepers need to be vigilant with their hive maintenance to keep AFB under control, by examining their hives regularly, and looking for any tell-tale signs of sickness.

Above left: Commercial beehives in a rapeseed field. Right: A hive wiped out by varroa.

STRAWBERRY FIELDS FOREVER?

In 2013 a Swedish study looked at the effects of pollination on the quality of strawberries. They studied fruit quantity, quality and value. Strawberries are a fragile fruit with 90 per cent of the harvest becoming unsellable after only four days due to fungal problems and fruit softening. (Who hasn't found a mushy strawberry in the bottom of the punnet?)

The study discovered that bee-pollinated fruit was heavier and suffered fewer deformities than hand-pollinated fruit, which therefore increased its commercial value. The strawberries were redder, had reduced sugar-to-acid ratios and were firmer, which improved the shelf life. According to the study, the longer shelf life reduced fruit

Strawberries are a delicate fruit prone to disease. Bee-pollination reduces problem harvests.

loss by at least 11 per cent. We do know that when a plant receives good pollination, hormonal growth regulators are produced, and it's thought that these are responsible for the improved changes to the fruit. This study emphasised, yet again, the importance of bees for successful food crops.

A 2014 survey to determine the risks to Australian crops, such as apples, cherries and blueberries, found that most growers depend on feral honeybee populations to pollinate their crops rather than commercial pollinators. The survey was conducted across a group of 100 growers who all understood the relationship between bees and pollination.

Worryingly, however, the growers surveyed weren't really considering the welfare of their pollinators and just expected that pollination would happen. This is all well and good—as long as varroa doesn't arrive, or there's a major natural disaster around the corner, like a bushfire.

Some growers in the survey also owned their own beehives, but like a plumber with a leaky tap, those hives were not a top priority when it came to farm maintenance, making them prime candidates for the spread of AFB.

Surveys like these reveal just how vulnerable Australian honeybee populations are, but also how lucky. In Australia we are, compared to the rest of the world, relatively disease-free. While varroa and CCD have not hit our shores, and AFB levels are low, we can't afford to be complacent. The risk is too great.

Most growers depend on feral honeybee populations to pollinate their crops ... This is all well and good—as long as varroa doesn't arrive.

Australia has so far escaped the global spread of varroa, which has caused huge declines in feral honeybee populations in other countries.

Pollen baskets

All bees collect pollen for protein, and they carry it in an amazing variety of ways. The European honeybee in this picture has mixed the pollen with a little bit of nectar and stuck it on a hair in an indent on its leg, called a corbicula.

Other bees don't have the indent, but rather a mass of hairs called a scopa.

Bees that have these special baskets comb off pollen from their bodies and pack it into balls for transport. Other bees, such as leafcutter bees, store the pollen on their bellies in a similar mass of hairs. The pollen in these baskets is not actually used for pollination, but delivered back to the hive. It's the residual pollen that's left sticking to their hairs while they forage between flowers, that results in pollination.

While bees don't have a problem with weeds, we do. Wild onion, or onion weed, is regarded as an environmental weed in Australia and is often treated with chemicals to help eradicate it.

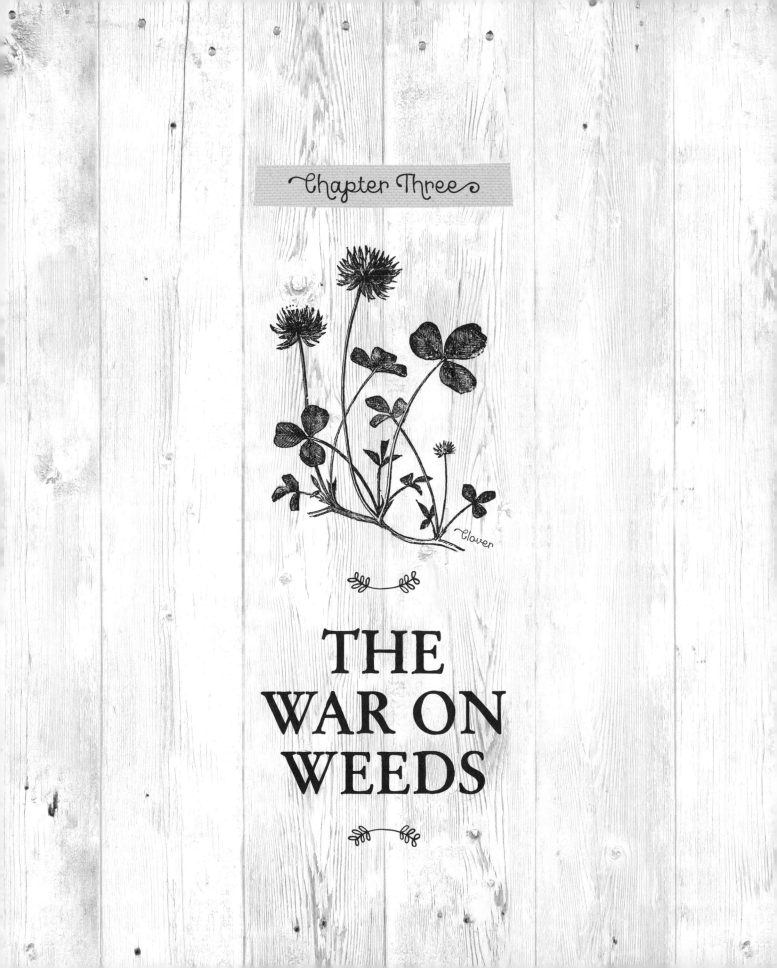

Chapter Three

Clover

THE WAR ON WEEDS

We all love a good bit of green grass, especially if it's as level as a bowling green and weed-free. When new development regulations stipulate that some 'green space' is required, that often ends up taking the form of a grassed area or, if it's an architect-designed garden, then so-called 'structural' and 'architectural' plants are preferred.

If you're a council planner then you love green canopy trees like plane trees, which are considered perfect because they're deciduous, fast-growing, pollution tolerant—and they produce an amazing green awning over our streets. They also produce fibres that block up all sorts of drains and irritate lungs, eyes and noses. Not only that, but from an insect's point of view the pollen they produce is of no use, and they produce little to no nectar.

All of those non-insect-habitat plantings unfortunately make up a large part of our local environment, which is pretty much the equivalent of a vast desert for our pollinators and other beneficial insects. The problem is that none of these modern options produce much food, and where Australia used to be nectar rich, our urban areas are becoming nectar scarce. In my local area, our honey

flows are decreasing and my neighbour George, a beekeeper and honey mead maker, has noticed a steady decline over the last 10 years; to the point where he gets precious little of the light-coloured honey he prefers for making mead and for that matter, little of any other colour, either.

My own experience with the hive I keep on my roof shows a huge decline in honey yields that's not slowing; and it's not because there are more hives and so more competition (which would be lovely if it were true), but because people quickly chop down perfectly good flowering trees without thinking about the consequences as new housing developments and infrastructure goes in.

While habitat loss is a huge problem for insects and other animals—a virtual war of attrition—the use of pesticides is turning the battle into a chemical war zone as well. Government bodies, farmers and local communities are all engaged in the war on weeds, and bees are just part of the collateral damage.

WHERE HAVE ALL THE FLOWERS GONE?

Even when people are doing good for the environment, they're not always thinking of the insects and other lesser known fauna. For instance, preserving remnant bushland is an important activity we undertake all over Australia, but often herbicides or other methods are used to

Lawns are often maintained using harmful chemicals—making them dangerous to birds and insects.

Some urban rooftop beehives are experiencing a steady fall in honey yields as city areas become scarce in nectar.

Parks and playing fields are regularly treated with herbicides to ward off flowering weeds—and the bees they attract.

Even weeds play their part. There's such a delicate balance between removing them all, turning the landscape into a wasteland, and leaving them there to continue their invasion.

eradicate weeds as part of bushland maintenance. Introduced plants are universally destroyed to create space for our native species and dirt banks are topped with mulch to prevent weed growth. But these activities actually have a detrimental effect, reducing the ability of the landscape to regenerate.

Take lantana. This much-hated, incredibly invasive noxious weed has come to provide a habitat for native bees in its hollow stems. Recent studies also suggest that lantana provides habitat for native rats and marsupials, as its spiny undergrowth is excellent shelter from predators. But in our haste to rehabilitate the landscape,

lantana is removed and burnt, and in the process many native bees are killed and animals left without their protective cover. And the impact could so easily be reduced; if only the workers chopped off the end of the pithy stems, bundled them up and hung them in a tree. While this doesn't help the animals who had used it as a shelter, at least bees could be left undisturbed and emerge next season to continue their work.

Dirt banks and bare ground are also an important habitat for many native bee species, such as blue-banded bees, teddy bear bees (*Amegilla bombiformis*) and other burrowing bees. If these areas are mulched, this obstructs their habitat.

Even weeds play their part. There's such a delicate balance between removing them all, turning the landscape into a wasteland, and leaving them there to continue their invasion. New research suggests that perhaps we shouldn't rush to remove some of those exotic weeds until we understand the relationship they have with the environment—we could be doing more damage than good.

Rather than rushing in to eradicate weeds we need to think about the role they play. Blue billygoat weed provides a great food source for solitary bees, honeybees, hoverflies, beetles and butterflies.

Councils that keep bushland areas regularly mowed and weed free using chemicals often cause harmful knock-on effects for bees and other animals.

DEAD ZONE AHEAD

One of the apiaries I look after is located in a remnant bushland area. Until recently, when I started complaining to council, the hives had a 10-metre-wide dead zone around them that was constantly mowed and sprayed with herbicide to 'help' the area remain free of weeds. What was once a lush belt of weeds and other non-native plants and trees for bees to forage, was being turned into an ugly wasteland of dead plants and brown grass. The area also contains a lagoon, so goodness knows what all this spray was doing to the wildlife.

Nearby there are many hectares of land that have been cleared, ready for construction of new buildings. Previously it was host to a small commercial apiary, but it's now been empty for a few months. When it was cleared all the trees were chipped and the entire area mulched with the chipped trees. It's beyond my comprehension why this was done. It could have been left to grow weeds or seeded with native flowers (which are hardly going to halt construction), but instead it's now another barren lunar landscape.

Meanwhile, councils across the country routinely spray with herbicides to keep weeds at bay, killing much more than just the weed in the process. One local council decided that the possibility of somebody getting stung by a bee in a playing field (and potentially having a fatal allergic reaction) was so great, they started a campaign to spray all the clover (*Trifolium* spp.) in their fields with herbicide. This to me seems a crazy over-reaction, and a demonstration of the lack of knowledge about the risk of bees to humans. If they'd only done their research before reaching for the chemicals they'd have unearthed this fact—in the 22 years between 1960 and 1981, no recorded bee-sting deaths occurred in the age group six to 19 years (this is the age group in which bee-sting

anaphylaxis is particularly common) according to Australian Bureau of Statistics data. This old statistic is another example of how bees have until recently just slipped through the cracks without much attention being paid to them (except for the scary, unsubstantiated kind).

Often I'll walk past a playing field or a park with a warning sign recommending you keep clear of the grass for 'x' number of hours because they've just sprayed the area with herbicides. When I've approached councils about the impact this is having on animals, they quickly refer me to the manufacturer's website, where it's claimed that the herbicide is safe for all sorts of insects and animals. Personally, I don't see why the grass needs to be clover-free, what's wrong with a few weeds?

Often herbicides need to be sprayed during certain times of the day to be effective and that time also corresponds with the forage time of bees, and if whatever councils are spraying needs to come with a warning for people to 'keep off the grass' then it can't be good for any creature.

CHEMICALS IN OUR MIDST

For years, timber treated with copper chrome arsenate (CCA) was used extensively in gardens, even in children's outdoor furniture and vegetable gardens. Many of us, or our children, would have played on furniture made of the distinctive green logs. Today, we realise that these timbers leach arsenic and other chemicals into the soil and into the hands of those playing on them or eating at the tables made of them. To think that the barbecued chicken eaten at a family picnic could have contained arsenic from your hands is a scary proposition that doesn't bear thinking about today. These timbers are being phased out and replaced by non-toxic versions, and it's now illegal to use CCA-treated timber in many places, including Australia.

This inner-city community vegetable patch takes an old-style approach to pest control that our grandparents used to use. Today we call it 'organic'.

There is a very long list of things we have commonly used in the past that we eventually realised were doing us harm. Chemicals like DDT—an insecticide that was found to cause cancer and damage wildlife populations; polychlorinated biphenyl (PCB), a synthetic compound used in electrics which caused huge environmental contamination; and the building material, asbestos, which contains thin fibrous crystals that can cause serious and fatal illnesses when inhaled. There is even an inner-city suburb near where I live that was home to a uranium smelter 100 years ago and still has soil contaminated by radioactivity. History is littered with these sorts of stories.

I wonder how many of today's frequently used garden herbicides, pesticides and fungicides, and all those 'heavy-duty' cleaning products that lurk under our sinks will be phased out in the years to come as we realise just how damaging they are to our environment and our health. And just how long they will remain in the environment and our bodies.

THE RISE OF THE 'CIDES'

In my beekeeping book, *Backyard Bees*, I mention my obsession with food, which sprouted from my Italian family's love of freshly grown produce and handmade smallgoods and cheeses. My childhood memories centre on my grandparents' acre block and its rich soil where they grew their own produce. What I also remember is the lack of powerful insecticides and herbicides. Yet somehow my grandparents still managed to produce amazing crops of vegetables and salad greens of all types.

Weeds were a problem that was managed by pulling them out, a tradition that continued long after the plot was no longer used to grow food. Whipper snippers hadn't been invented and we couldn't conceive of using the lawnmower to chop up and mulch weeds in case we damaged it. (It wasn't until I was given an old broken lawnmower that we could afford to lose, that any mechanical weed eradication was considered. Subsequently, I fixed the lawnmower and it worked a treat at chopping the weeds off and mulching at the same time, much to the surprise of those nay-sayers in my family who'd been weeding by hand for years.) We'd never have imagined it, but these days our normal way of doing things comes with fancy terms and labels, like 'organic' and 'sustainable'. How trendy these old ways have become.

Today, the average backyard vegetable grower might consider using a herbicide to eradicate the

Take no prisoners. This fruit tree is being sprayed with a pesticide.

> *When bees take fungicide-infected pollen back to their nest, the result is a sicker population.*

weeds before planting some lettuce because we all feel so time poor and it's much quicker and easier than pulling all the weeds out by hand. After all, the label says it's okay to use and doesn't harm bees. The same backyard gardener might use a whole assortment of chemical fertilisers, pesticides, herbicides and fungicides (or simply 'cides', as I like to call them—'cide' meaning 'killer'), all the while trusting the manufacturer to disclose what these chemicals are doing to the natural environment.

In an effort to reduce farm labour, many farmers are now mixing various pesticides and fungicides together in tanks so they only need to go round their fields once with the spray applicator. This can have unfortunate side effects, with tank mixing being the apparent cause of huge bee losses in the almond fields in California in 2014. In that instance, approximately 80,000 hives were lost—because the chemicals had been mixed together they became extra toxic to bees.

Recent studies show that some of the routine seed treatments that our farmers use are ineffective. We even have herbicide-resistant weeds appearing. Weedscience.org reports that in 2015, herbicide-resistant weeds had been reported in 86 crops in 66 countries. Resistance was reported in a staggering 157 different herbicides. In Australia, we have had herbicide resistance reported as far back as 1982 and yet we are still using herbicides.

We are constantly finding out that 'cides' we were told were safe are not, and are instead harming our environment. Take fungicides, for example. For some time they were considered safe, based on the simple observation that when bees foraged on flowers that had been sprayed with fungicides, they didn't drop dead. (That was in contrast to what would happen if they foraged on

Herbicides are usually sprayed during forage times for birds, bees and other insects.

flowers that had been sprayed with an insecticide, in which case they did die.) Two different studies were conducted that looked at hive health and also native pollinators' health. It turns out that when both types of bee take fungicide-infected pollen back to their nest, the result is a sicker population that's more susceptible to fungal infections. As we know, bees collect pollen and store it for their brood. That pollen contains many microbes beneficial to bees' health, and it's assumed that the residual fungicide kills off those useful organisms.

THE MICHIGAN EXPERIMENT

Recent research indicates that a scarcity of flowers leads to a greater prevalence of pests and disease

A diversity of food types reduces the risk of illnesses being introduced into local bee populations. This garden bed provides a rich smorgasbord for bees and other beneficial bugs.

Imagine eating the same meal every day. Apart from it being boring, your health would suffer from not having a well-rounded diet. Perhaps a lack of floral diversity is also having an effect on our bees.

as more bees and insects are forced to visit the same flower, and they leave behind traces of any ailments they're potentially suffering. It's a bit like a dirty handrail in a bathroom spreading disease; bees become more at risk of developing illnesses due to such close foraging. In the past, when there was an abundance of food, these ailments were not as easy to spread—I suspect this lack of forage could be part of the pollinator loss puzzle. A lack of quality forage could also be bad for insect health. Imagine eating the same meal every day. Apart from it being boring, your health would suffer from not having a well-rounded diet. Perhaps a lack of floral diversity is also having an effect on our bees.

In Michigan, USA, an experiment was conducted in blueberry pollination. Researchers Brett Blaauw and Rufus Isaacs seeded fields with 15 varieties of perennial wildflowers and then observed bee activity in those fields and the adjacent blueberry fields. They were looking at the effect the loss of habitat was having on local native bee populations and whether those native bees could replace the threatened European honeybee for blueberry pollination. The results were astounding. The researchers found that after two years the native bee numbers had increased so greatly, and the extra pollinators were increasing the blueberry crop yield to such an extent, that the cost of planting the wildflowers could be recouped from the extra yield. According to the study, a two-acre field planted with wildflowers adjacent to a 10-acre field of blueberries boosted yields by 10 to 20 per cent.

This is important in many ways. We are often tempted to dismiss native bees in our agricultural pollination, but maybe a rethink is needed, and just maybe all those years of mowing road verges and using herbicides to keep the weeds down have actually been costing farmers money. If they left roadside weeds alone and let them flower, not only would they save the time taken to mow in the first place, but they would get increased yield because the extra habitat would encourage native pollinators.

SAVING BEES GETS MAINSTREAM

Syngenta, a large worldwide agricultural company, runs a program called Operation Pollinator, where it encourages wildflower planting on all sorts of land, including golf courses. Their campaign attempts to educate farmers in the use of well-chosen wildflowers to retain soil and assist beneficial insects, where once they might have kept areas around their crops bare for easy access. All these changes herald a new understanding in the relationship farmers have, or should have, with insects.

Meanwhile, the US government is officially reassessing the role of herbicides and has created a Pollinator Health Task Force, as discussed in chapter one. The aim is to rehabilitate seven million acres of land as pollinator habitat in an effort to save bees and butterflies, and to recognise their vital role in food production. There's even a pollinator garden planted at the White House alongside the now substantial kitchen garden, replete with a beehive producing White House honey, which is then bottled in an ornate jar.

Worker bees live for about four to five weeks. During this short time they're constantly at work, and produce about a quater of a teaspoon of honey.

Chapter Four

Leafcutter Bee

THE WORLD OF BEES

Depending on who you believe, there are between 20,000 and 25,000 bee species in the world, with many more that probably exist and haven't been discovered yet. Amazingly, bees exist on every continent except Antarctica and can be found anywhere that insect-pollinated flowering plants exist. That's a lot of bees and a lot of ground to cover.

All bees are part of an insect family called Hymenoptera, which also includes wasps, ants and sawflies. It's believed that bees evolved from wasps about 130 million years ago, around the same time as the first flowering plants, called angiosperms. The oldest specimen of a bee is a fossil called *Trigona prisca* that was found in some amber. It's believed to be 74 million years old, from the Cretaceous period.

Some bees are tiny little things—the smallest is thought to be *Trigona minima*, a stingless bee whose workers are about 2.1 mm long, and the largest *Megachile pluto*, a leafcutter bee whose females can reach a length of 39 mm.

THE BUZZ ABOUT EUROPEAN HONEYBEES

When we talk about bees, the one that gets the most PR is the European honeybee (*Apis mellifera*). This species of bee lives in large hives of 60,000 to 80,000 bees, which makes them an excellent pollinator—huge numbers mean lots of hard-working foragers. This workhorse pollinator was brought to Australia by Captain John Wallis on the convict ship *Isabella* in 1822. (By a ridiculous coincidence, he had also been master of an earlier ship named *The Three Bees*.) Wallis successfully imported seven hives, two of which were bought by D'Arcy Wentworth, the then owner of Vaucluse House in Sydney (which now happens to have an apiary managed by my bee business, The Urban Beehive). These pioneering bees fared very well and were well suited to life in Australia, with many swarms invading the countryside and establishing hives. In fact, a newspaper of the time, *The Monitor*, reports a few years later that if Wentworth had collected all his bee swarms he might have owned 70 hives instead of the seven into which his initial two had grown. Australia is such a nectar-rich country that the original hives just kept expanding by swarming, populating all the parts of the country where we have flowering plants. Nobody knows exactly how many beehives exist in Australia today, but they've become a naturalised part of the landscape, with hundreds of thousands of hives.

European honeybees are referred to as 'social bees' because they live in a hive structure with a queen and worker bees—essentially a society. The beehive operates pretty much as a single entity, something we call a 'super organism' where all the individuals work together doing different tasks, all designed to keep the hive operating efficiently. The hive collects pollen as protein to feed young developing bees (called brood), nectar as carbohydrate (which is stored as honey and mixed with the pollen for the young bees and eaten without pollen by the adult bees), and water, which is used in the hive's air-conditioning system.

Honeycomb

Honeycomb is the hexagonal storage compartments built by worker bees within a hive. It consists of hundreds of wax cells (made of secreted honey) and is used to raise brood, and to store honey and pollen. Once full, each cell is sealed with a layer of wax.

EUROPEAN HONEYBEE HIERARCHY

Inside a European honeybee hive there are three sorts of bee: the queen, drones and Worker bees. Every bee has a role to play and for queen bees that role is to be the hive's egg-laying machine. While the queen might be central to the whole life of the hive, she's not exactly in control: it's the worker bees who get together and vote on important decisions. Other bees' duties are coordinated around the queen's egg laying and the health of the hive.

~Queen bees~

Queen bees (see large bee, top right) are a marvel. They start life as a regular fertilised egg that would normally develop into a worker bee, except they're fed royal jelly for their whole development, turning them into a queen. (Royal jelly is a nutritional secretion from worker bees, generally given to newly hatched brood.) Queen bees mate at the beginning of their lives and will mate with a number of drones at that time. They then store and use the semen in the reverse order that they received it. When you open a beehive you may notice that some of the worker bees look a little different from each other, and that's the different semen causing slight variations in the bees. The queen bee emits a pheromone that's the natural scent of the hive and tells the bees that this is their home. Queens last as long as their semen stores remain viable, which can be from one to five years depending on how well they mated.

~Worker bees~

The worker bees are all female and do all the work around the hive, from cleaning and feeding to guard duty and foraging for nectar, pollen and water. They work until they literally wear themselves out and can't fly anymore. Their journey from birth (shown above) to death takes four to five weeks and during that time they each make about a quarter of a teaspoon of honey. Worker honeybees forage in a five- to eight-kilometre radius from the hive, covering a very large area. They're constantly on the lookout for good nectar sources. When they find one they return to the hive and, using a waggle dance, communicate the location of the nectar to the other bees and share a small taste of it so the other bees can find it.

The hive

Drones

The drones are the male bees, whose only job, it seems, is to find a queen and mate with her. They don't do any work around the hive—they can't even feed themselves; the worker bees do that for them. Drones come from unfertilised eggs that the queen deliberately lays for that purpose, and therefore have no fathers, just grandfathers. It might be a bit hard to get your head around but that's how it works. Drones are large, noisy bees with huge eyes and no sting. You can spot them easily in a hive if you don't hear them first. They spend their life—drones have a life span of about 40 days—congregating with other drones, looking out for the queen to mate with. When they see her they all take off trying to catch her.

The colony is very organised and everybody has tasks to do, the most important job being to maintain the hive temperature at around 35°C. The worker bees control the temperature in the hive by using their wing muscles to generate heat and cooling the hive through ventilation. By spraying water and evaporating it (a form of air conditioning), they also control the humidity to within strict levels necessary to keep the brood and queen healthy. Bee colonies are a regenerating organism that will last indefinitely unless a catastrophe occurs that kills the queen and no new queen is raised, or she dies while mating due to a predator such as a bird or is otherwise lost.

Inside the hive there are distinct levels. Generally, at the top of the hive there will be honey stored in an arch and recognisable by its white cap; the white cap is wax that the bees have used to close up honeycomb cells full of ripened dehydrated nectar that's now honey. Under the honey there is often an arch of stored pollen mixed with honey, often called 'bee bread' and easily recognised by its multiple colours. Under all the food is the brood, or developing bees. Here you'll find various stages of bee development, from eggs to larvae to 'capped brood'—the last stage of development and easily recognised by its tan or light brown velvety breathable cap.

To maintain a healthy hive (and produce honey), honeybees need good quality, high-protein pollen, ample nectar and a water source.

European honeybees are the bees responsible for pretty much all the honey you put on your toast, with a single hive producing in excess of 50 kg of honey per year and in a well-managed hive sometimes twice or three times that—enough for a lot of crumpets.

European honeybees are the bees responsible for most of the world's honey production. A typical hive can contain between 60,000 and 80,000 bees and produce in excess of 50 kg of honey per year.

NATIVE BEE SPECIES

In Australia we have over 2000 species of native bees, and new ones are being described all the time. In fact, as recently as 2013 the tetragonula bee was renamed as part of a reclassification of what many call 'stingless' bees. (It was previously called trigona—indicating a different group of bees.) Of those 2000 species, many can, despite assurances to the contrary, deliver a sting and many have barbless stingers, which means they don't die when they sting you, so they can sting multiple times. Many people don't feel the same fear of native bees that they do of European honeybees, and native bees in general seem to be thought of as rather cute. Many are indeed gorgeous looking with all the colours of the rainbow represented. (I think all bees are cute, but that's just me being a beek, as I like to call us bee geeks.)

Our native bees range from about 2.5 to 25 mm long and have many unique features. Some rely on a single flower species for their food (with the flower, in return, relying on that bee for its survival). Others are more opportunistic and forage on what they can find. Surprisingly, many native bee species can be found in the average backyard and are often overlooked or mistaken for flies. They don't produce much honey for humans, but they do fulfil other essential roles.

This section contains a few of my favourite native bees. These are only a tiny selection from the thousands of bees that can be found across the continent. There are many, many more that are probably visiting your backyard without you even noticing, and if you keep on looking, who knows, you might even spot one that hasn't been seen in your area before.

Many native bee species can be found in the average backyard and are often overlooked or mistaken for flies. They don't produce much honey for humans, but they do fulfil other essential roles.

~Teddy bear bee~

Now here's a bee that would give Paddington Bear a run for his money in the cuteness stakes. It's one of my favourite Australian native bees and it's the most adorable bee ever. The aptly named Teddy bear bee (*Amegilla bombiformis*) is a lumbering, tan-coloured, furry leviathan of a bee. It's comparatively huge at up to 15 mm long and though it's not as common in urban Australia as some of the other bees, I've seen it near where I live in a very built-up area, foraging on mock orange (*Murraya paniculata*). This is a great example of how our natives bees seem to find homes in the most remarkable of places.

Teddy bear bees make a nest consisting of several bucket-shaped cells at the end of a burrow up to 10 cm long in creek banks or garden rubble. Because the teddy bear bee makes its nest in soil, it's at risk in urban areas because, tragically, its habitat is being razed by developers who remove open soil or concrete over it. I wish more people could see this bee in action; if they did they would make sure they maintained its habitat for sure.

Stingless bees that can bite

We have a couple of native bee species—*Tetragonula carbonaria, Tetragonula hockingsi* and *Austroplebeia australis*—that live in a social structure like European honeybees. These natives are commonly referred to as stingless bees because they cannot sting. They have strong mandibles and can bite, although no human would succumb to such an attack—it's more annoying than painful. In the warmer parts of Australia these bees can be found in many a backyard or community garden.

These small black bees are about 4 mm long and live in colonies containing a few thousand bees. They really do look a little bit like a fly until you get closer, and then the differences become apparent, especially when they have full pollen baskets (as shown here).

One easy way to tell the difference between a fly and a bee is to look at the wings. Bees have four wings and at rest they're usually folded shut. A fly only has two wings and they're always open. Tetragonula bees live in the warm parts of New South Wales, Queensland and the Northern Territory, usually in tree hollows, but increasingly, as we become more native-bee aware, in small, white, man-made hives about the size of a toaster. The exception is in far northern Australia, where they're often found living in urban infrastructure and are sometimes considered dangerous. As a consequence, they're often exterminated, which I think is quite sad.

Unlike the larger European honeybee, stingless bees do not thermoregulate their hives, which makes them susceptible to extremes of heat and cold.

Researchers are still working to understand what these bees actually pollinate, but it was established in recent research that these bees visited some endangered species of native plants, and it stands to reason that being native they would be pre-disposed to local native varieties. Yet I've seen them visiting basil, pumpkins and other vegetables. In my own inner-city backyard I have a hive of *Tetragonula carbonaria* (which are better at tolerating the cold in my area than *hockingsi or Austroplebeia*), but unfortunately I've never seen them visit my purple-flowered borage, which I planted for their benefit and which is loved by bees the world over. Despite it being right outside the hive and in full flower, they zoom off elsewhere to some other more favoured variety of plant.

The small stingless bees like tetragonula are often used for commercial pollination of macadamias, and until recently were also thought to be pollinators of mangoes; that's until scientists realised that a number of fly species were doing more mango pollination. That's just another example of the interwoven roles that all insects and plants play.

Tetragonula bees are a little tricky to spot. They're only 4 mm long, about the same size as a fruit fly.

Blue-banded bee

One common backyard bee that could never be confused with a fly is the blue-banded bee (*Amegilla cingulata*). This charming blue-striped creature is a common visitor to backyards right across Australia. It's about the size of a honeybee, but that's where the comparison stops as the honeybee's yellow bits have been replaced with deep blue. It's a bee that's hard to miss.

Unlike the Tetragonula bees, blue-banded bees have stingers. But like most bees, they're unlikely to use them unless you accidentally step on them or really get in their space. Blue-banded bees are referred to as 'solitary bees' because they don't live in social colonies with a queen and her workers bees and drones. Instead, they live alone or in aggregations where many individual female bees set up home in close proximity to each other; a bit like a block of flats except the idea is that grouping together gives protection from predators—safety in numbers. I guess you'd want to be the one with your home in the middle of the group in case your neighbours started being picked off by birds or other predators.

Blue-banded bees like nesting in soft clay soil and are often seen under houses in Queensland or boring into soft mortar in New South Wales, much to the concern of some householders. In reality, they do little damage to the building—white ants they most definitely are not!

This bee generally builds its nest as a kind of tube in the soil, in a shape that resembles a section of bamboo or cane, with each individual chamber containing an egg and stores of nectar and pollen. The male bees are nowhere to be seen, preferring to hang out together and even sleep together on twiggy stems of plants using their mouth parts to hang on while they sleep.

Blue-banded bees sound different from honeybees when they fly, having a higher pitched buzz that gives a hint as to one of their unique abilities: buzz pollination. These bees grab hold of the stamen of plants that are members of the deadly nightshade family, like tomatoes, potatoes and eggplants, and use their wing muscles to transmit vibration through their head to the plant, releasing the locked-in pollen (see page 38 for more on buzz pollination).

Another common native bee is the leafcutter bee (*Megachile* spp.). Those semi-circular cuts to your rose bush leaves are not from a caterpillar or your kids playing around with nail scissors, but a bee that cuts out arcs of leaf and flies home with them to line her nest. The plant suffers no injury and the bee gets a home. When these bees head home carrying their leaf slice, I always liken it to the guy in the tiny car you see driving up the street with a king-sized mattress on the roof. Somehow, the leafcutter bees manage to get the bits of leaf home, roll them up and poke them into the nest hole.

Not surprisingly, these bees prefer soft leaves (and roses seem to be a favourite), though they may also try butterfly bush (*Buddleja* spp.) and wisteria leaves. This bee is solitary and lives only several weeks in the summer. Like many solitary native bees, the mother bees never see their offspring hatch, as she has perished before they develop. She builds the nest, laying a few eggs separated by a membrane, a bit like a cross section of bamboo.

There are 150 described species of leafcutter bee in Australia, which gives you an idea of just how many specialised bees we have. They range from 5 to 14 mm in length and make their nests in hollow tubes, sometimes old borer holes in rotting timber or bee hotels (see chapter 10).

Once again there is not a lot of knowledge about exactly what these bees pollinate, but I've seen them buzzing around my tomatoes, basil and rocket flowers and it's thought they pollinate clover, alfalfa, tree fruits, onions and carrots.

In the 1930s, a European leafcutter bee was introduced to the USA and is now a very important pollinator of alfalfa for seed production. The pollination process is managed by specialists, and, as a result, alfalfa seed yields have increased dramatically compared to pollination with honeybees or no bee at all. With the leafcutter bee pollination method, 40,000 to 60,000 bees are used per acre and the management is such that hatching is synchronised to match the alfalfa flower bloom. Apiarists do this by placing special habitat in the alfalfa fields and then removing the nests to cold storage, then incubators, and finally taking them back to the alfalfa fields when it's time to hatch. The cold temperature delays the hatching of the bee until the crop is ready for them.

It's quite an industry, with all sorts of intervention used to keep the bees healthy and free of disease like chalkbrood (a common fungal infection of honeybees). X-ray machines are even used to count the number of developing bees so farmers know whether they need to purchase extra stock. It all sounds a bit mechanical to me and perhaps a bit sad for the bee to be so commercialised.

Metallic carpenter bee

The golden-green carpenter bee (*Xylocopa aeratus*), gets its name from the metallic-green colour of the females and its nesting preference which is a hollowed-out tube about 30 cm long in grass trees (*Xanthorrhoea* spp.), or other soft wood such as paperbark trees (*Melaleuca* spp.). (The males have fine yellow hair on their bodies.)

It makes a low-pitched sound when flying and is quite a large bee. This native bee and many others rely on old, sometimes even dead trees for habitat and are a very good reason for not clearing away old timber when rehabilitating bushland areas or, at times, even your own backyard.

NOT ALL BEES
ARE GOOD BEES

Australia has many introduced species of animals and insects—some brought in to do a job that they ended up not doing very well and instead becoming a pest and a huge source of regret. An example of this is the cane toad, brought in from Hawaii to control native cane beetles, and becoming out of control themselves. Some bees are introduced unintentionally, hitching a ride on a recreational boat or container ship, but they all tend to become a problem. These feral bees often favour non-native plants when they forage, which then provides a pollinating boost to weeds and other exotic plants, which then become so successful they quickly get termed weeds too. This then tips the scales in plant and animal diversity in certain areas, as the introduced bees out-compete the local bee species for nectar and pollen resources.

Carder bee

The carder bee (*Bombus pascuorum*) is a type of bumblebee, which is spreading across the east coast of Australia. The carder bee is sometimes called the 'meter box carder' because they often establish their fluffy nests in meter boxes. The carder part of their name comes from their habit of collecting plant fibres and carding them together like a cotton carder in a mill. The nest ends up looking like cotton wool. Apart from meter boxes, they like nesting in window frames.

Originally from South Africa, this bee was first seen in Brisbane in 2000. Recently it was seen in Victoria, so has spent no time at all making its way down the coast. Humans are responsible for transporting it in building materials and even on the window sills of caravans.

We really don't know what effect this bee will have on our local biodiversity, but I guess over time we will see what it chooses to forage on, and if it boosts a weed that's currently under control.

Bumblebee

In Tasmania, you'll find an introduced bumblebee (*Bombus terrestris*), a large social bee native to Europe. They live in social nests in the ground with a queen, workers and drones, and colony sizes usually vary between 500 and 1000 bees. These bees have formed a large feral population on the island after arriving accidentally or being illegally introduced in 1992. While it's a cute bee—furry and fat—it's a serious threat to Tasmanian biodiversity as its large muscle mass means it can warm up and get going way before the other bees get their engines started on cold Tasmanian days. I once read a story about a bee researcher in the UK trying to humanely dispose of a bumblebee nest, only to find that after four days in the freezer they were still alive: a truly amazing feat.

This bee does seem to be having an impact on introduced plants, such as common rhododendron in Tasmania, which has the potential to become a noxious weed (it's already classified as one in parts of Europe). This bumblebee also has a habit of chewing flowers and bypassing the pollen collection, meaning they're not pollinating, just consuming the reserves that have been left for the pollinators.

Asian honeybee

The Asian honeybee (*Apis cerana*) was first detected in Cairns in 2007. It's a common bee across much of Asia and the Pacific islands and it's believed to have arrived here in a boat mast. It's a very similar looking bee to a regular honeybee, with yellow and black stripes, but it's about 5 mm shorter and the stripes seem more regular than those on the honeybees we're used to. The Asian honeybee is spreading rapidly south, which it does by hitchhiking as a swarm, and swarming is something it does a lot, often settling on things like trucks or truck loads.

Recently a swarm hitchhiked on a campervan being towed interstate from Cairns to Darwin, causing a desperate search for the queen bee that was missing from the swarm when it was detected. If it gets established in another state there may be no stopping it. *Apis cerana* steals honey from existing European honeybee hives and probably prefers to pollinate certain flowers—we just don't know yet what impact it will have in that area. The biggest concern is that it's a natural host for the devastating varroa mite and having it established means that should varroa arrive it will have a host ready to go.

Lots of work is still being done to try to reduce the Asian honeybee's spread across the eastern states, but it seems to be slowly spreading regardless. If varroa established itself here in Australia it

Lots of work is still being done to try to reduce the Asian honeybee's spread across the eastern states, but it seems to be slowly spreading regardless.

would be devastating to our local agriculture (not to mention our bees), as our feral honeybee population would be severely impacted. Little is known about how the introduction of Asian honeybees might affect our plant mix, but it would be a potential disaster if they were found to prefer some introduced weed that has previously been kept in abeyance through low pollination.

Killer bees

Now you're probably wondering about killer bees, but don't worry, we don't have them here. Killer bees (also known as Africanised honeybees) are recognised by their extreme defensive behaviour, but interestingly they're very productive and when managed for honey production easily out-perform the regular honeybee.

They're a real phenomenon that came about through an attempt to breed more productive bees by crossing an aggressive African bee with the common European honeybee in the 1950s. This was all done in Brazil, and special screens were fitted to the beehives so the larger male drones and queen bees couldn't get out of the hives and mate with the local population. Unfortunately though (like a scene from *Jurassic Park*), a visiting beekeeper removed the screens and a number of swarms escaped into the wild. They arrived in the USA in 1985 and were found in California in 2015.

The term 'killer bee' really should stay in the realm of fiction, though, as they actually haven't killed that many people, although they have been known to chase a potential threat for 500 metres—so you would want to be wearing good running shoes if you came across them.

Bee swarms

A bee swarm is a marvel to behold. It's what happens when about half the bees leave a hive, along with the old queen, in a huge noisy cloud, only to settle temporarily nearby in a huge, brown cluster. Despite their menacing appearance en masse, bees are usually at their most passive when swarming, as they have no hive to protect and are instead looking for a place to found a new colony. Swarming commonly happens when spring starts, with the worker bees preparing a special cell where a new queen will be raised. Two weeks later, just before the new queen emerges, the old queen and her entourage prepare for the journey by gorging on honey. Then they leave the hive in the hope of finding a place to set up a new home, with bee scouts tasked with looking for good locations.

There are many environmentally safe ways to remove garden pests without having to reach for the spray. An old vacuum cleaner, for example, is a great way to get rid of stink bugs from citrus.

Chapter five

SIMPLE CHANGES ANYBODY CAN MAKE

Not everybody wants to become a beekeeper, I realise. But with just a few changes to your garden and your lifestyle you can make a massive difference to its attractiveness to bees and, as a result, to the biodiversity of your surrounds.

GET RID OF THE SPRAY

Once upon a time personal insect repellent was all you needed to ward off flies and mosquitoes and 'Avagoodweekend'. Not these days. Instead of using repellents to keep annoying bugs away during a backyard meal, many people instead pop into the supermarket for an automatic dispenser of insecticide that will annihilate them all. When you use one of the various available toxic mixes that automatically spray across your backyard, you need to think about what that might be doing to the rest of the insects.

To quote Isaac Newton, 'To every action there is always opposed an equal reaction'. In this instance, making your backyard an insect-free zone with a liberal dose of insecticide means the whole balance of the ecosystem is affected. Even birds won't come if there are no insects. Which is fine if you want to live in a world like the one depicted in an apocalyptic sci-fi movie, but I prefer my greens fresh and leafy.

So the place to start is under the kitchen sink, where you've got those insecticides stowed. Ask yourself whether you really need to use them. Do you really have an ant problem, spider problem or other creepy crawly problem in the backyard (or even inside the house)? And do you need to apply long-lasting insecticide to all surfaces near and far to have a happy and healthy life? I suggest the answer is no.

If you believe the advertising that bombards us daily, every surface in the home is a potential death trap dripping with germs or filthy insects. You even need to disinfect the toilet bowl, especially up under the rim! Because it might have germs, and goodness knows a toilet bowl with germs would be a terrible thing. I have friends who won't flush their toilet with rain-tank water because it's not clear and clean enough, preferring to use good-quality drinking water. I'm not sure how they'd fare in that sci-fi apocalypse.

DO A STOCKTAKE

Take a good, long, hard look at your household chemicals and cleaning products: you'll be surprised at just how many there are if you line them all up. Start by placing them together on a table and looking at the contents. Some of them are really bad for you; even garden sprays that hint at being organic or natural are mostly not, so read the label carefully. An excellent example of this is pyrethrum, which is a powerful insecticide that comes from a chrysanthemum— so it must be pretty harmless, right? Not if you're an insect. It's a very toxic substance that's a repellent in low doses, but fatal to insects in high amounts. It's also toxic to humans, and testing reportedly suggests that if you consumed one per cent of your body weight of the stuff it would kill

Making your backyard an insect-free zone with a liberal dose of insecticide means the whole balance of the ecosystem in your backyard is affected.

Check the source and contents of potting mix and plants when you purchase them. Sometimes they're host to dormant weeds and chemicals that you don't want to introduce into your garden.

you. (Just because it's often naturally extracted from flowers doesn't mean it's harmless.)

I won't list the names of potential insecticides here—there are a lot and the names change frequently. But I suggest doing an internet search for each product and looking at the safety sheet for it. My rule of thumb is, the more precautions attached to it, the more toxic it is.

And remember that chemicals are not just found under the kitchen sink or in the garage. As gardeners we often use chemical fertilisers when good old blood and bone would be just as effective. You need to eliminate as many chemicals as you can, including chemical fertilisers. Go back to the basics and use worm castings and animal manure as fertiliser. It's very effective and manure is of course how plants are meant to be fertilised. Use organic where you can to eliminate chemicals in the poo or other garden products, and beware of products that are not certified organic or not bearing the standards mark, as they may not be good mixes. I once ordered 10 cubic metres of garden mix for my new vegetable gardens only to have a highly acidic mix containing a lot of ash from a fire turn up, which took heaps of garden lime and organic matter to balance out to something in which I could grow my veggies.

Take stock of all the chemicals you use and eliminate the ones you don't need. The environment will thank you for it and so will our bees and other insects.

ASK QUESTIONS ABOUT ALTERNATIVES

The internet is awash with alternatives to household chemicals, but some of these alternatives, while effective, are also environmentally damaging. For example, there is a recipe doing the rounds for weed killer that uses a large proportion of salt and some vinegar,

Boiling water will kill weeds just as effectively as chemicals.

and this mix is a cracker—it will kill the weeds for sure. But hang on a minute. The ritual of spreading salt on fields in conquered cities to symbolise a curse on their reinhabitation is part of historical lore, and many farmers know all too well that salt is the enemy of plants. So while this weed killer will get rid of your weeds, it will also inhibit all the good growth, and unless you want to be a modern-day Lawrence of Arabia or really want to replace your lawnmower with a dune buggy, you probably don't want a desert in your backyard.

Many weeds can be killed using alternative means, such as fire or boiling water, or even covering them with newspaper, until they use up all their stored energy and perish, which is very useful with some nasty weeds. One such weed is onion grass (*Romulea rosea*)—it can't be removed by hand as it spawns hundreds of replicates by exploding little bulbs as they're pulled out of the ground.

Be aware when spraying plants, especially in the colder months, as wet fur can be fatal for bees.

Which brings me to another question. My garden beds used to be onion grass free, but now the insidious weeds are everywhere, along with clover. The question is, where did they come from?

And the answer is, with your plants. There are many hidden problems in your local nursery, where you would expect to be able to buy healthy plants in clean soil for your new bee-friendly garden. But even at nurseries you have to ask questions. Quite apart from giving you a dose of weeds with your purchase, often nurseries treat their plants with systemic insecticides. This makes it easy for the nursery or hardware chain to keep nice-looking plants, but it's no good for the insects—the entire plant becomes toxic from the leaves to the nectar and pollen. They may also use other herbicides to keep weeds at bay until you've actually made your purchase, which is great until you get the plants home and a few weeks later you have onion grass sprouting up everywhere as I do.

So ask about the potting mix when you're purchasing plants and if you don't get a good, clear answer, seek out a nursery that will give you full satisfaction. Also, be very suspicious of anything that purports to be bee-friendly, especially any chemical product, as it's unlikely to be so and more likely to be just good marketing. There was an example in Europe where a spray that was toxic to bees was packaged with seed to grow bee-friendly flowers—a very cynical marketing exercise indeed.

You need to be able to see through the marketing hype that now surrounds all things to do with bees. The only way to be sure is to ask questions, lots of questions.

BUT I REALLY NEED TO SPRAY

If you really do need to spray, here are some thoughts about best times to do this, though at the risk of becoming repetitive, I do recommend you avoid spraying at all costs.

Commercial growers are beginning to understand that they need to think about what and when they spray. In the professional pollinator industry, many beekeepers complain about having their hired-out bees sprayed by the farmer, who then complains that the bees are not pollinating crops as well as they should. It's a bit hard to work when you're dead! Fortunately, farmers are being educated now about the best practices.

I need to stop for a moment here and explain what it's like for bees to fly. It's a hard task. They're not the most aerodynamic of insects and when the ambient temperature drops below about 13 degrees Celsius, or it's really windy, rainy or heavy with dew or fog, then they stay at home. If they're out flying when the temperature begins to drop they're able to generate their own body heat, but if the ambient temperature falls below 10 degrees they will eventually be unable to fly and become comatose on the ground, probably dying from exposure. (Cold nectar can also chill a bee and cause the same effect.) Sometimes if you see a bee like this and pick her up your body warmth is enough to get her flying again.

So getting back to the point—if you're out in the garden and happen to spray a bee, even if it's a relatively harmless substance, it will weigh them down and may cause them to be unable to fly. This is no problem if it's warm, as they can wait for their wings to dry, but if it's down around 10 degrees it can kill them. And of course if the spray is really toxic, dry wings are not going to help very much either. So if you really need to spray, avoid blanket spraying and try to use a nozzle with a single targeted stream when spraying insecticides.

Another consideration is that many sprays are much less toxic when dry. This could be because the active poisonous ingredient is not always the biggest threat to the bees; sometimes it's the surfactants—the substances that are added to the spray to help increase its ability to spread and wet the plants. Often the label on the container will explain that it's non-toxic for bees when it's the surfactants (which are not tested) that prove fatal.

Research on almond plants suggests that bees finish harvesting pollen by mid-afternoon, so if you wait until late afternoon or early evening when pollen collection has finished, you could be exposing fewer bees to the spray and therefore giving them a greater chance of survival.

There is really promising work being done by planting companion plants with crops and you can do this too. Companion plants add habitat and forage for natural predators so careful planting can reduce the need for pest control ... the reverse of past thinking. (I go into this a bit more later on. And if you're still determined to spray, we have a section on safe spraying solutions at the end of the book.)

INTEGRATED PEST MANAGEMENT (IPM)

This book is about building gardens for insects—mainly bees, but other insects as well. In your approach to pest control you need to consider the insect you are targeting. You don't want to rush into insect control like a soldier with a flamethrower—a more considered approach is best.

These days there is an acronym for everything and believe it or not, bug control is called integrated pest management (IPM). As the name suggests, it means considering all the options, which I will summarise as 'not just resorting to a gigantic can of spray'.

But once again, I want you to think for a moment. Your garden should be supporting all insects to maintain a healthy biodiversity, so is that cabbage moth (*Pieris rapae*) you might spot really out of control, or is it just that you prefer to look out at your garden and see lovely, untouched leaves? Basically, the garden is there for all bugs, good and bad, and it's only when they're truly out of control that you should be considering any remedies, even the organic ones. A few holes in leaves or a slightly chewed lettuce is no real problem. Not even a first world problem.

As I've mentioned in previous chapters, the overuse of pest control and our desire for perfect-looking plants is part of the reason for the pickle we find ourselves in at the moment. Or at least, the pickle bees and other insects find themselves in. Which will ultimately be our pickle.

So don't overdo it, let some of the so-called bad bugs survive as well as the good ones. It's all part of the food chain that we rely on heavily for our own sustenance. And before you reach for the control options listed here, read the section on good bugs in chapter 11, as there may be an alternative there that uses no chemicals—just the miracle of nature.

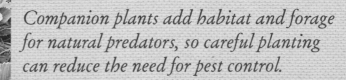

Companion plants add habitat and forage for natural predators, so careful planting can reduce the need for pest control.

The perfect bee-friendly garden has an array of different flowering plants, both native and exotic, that flower at different times throughout the year.

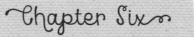
Chapter Six

LET'S GET
GARDENING

Next is the fun bit—building your insect paradise, right in your own backyard. Have a look in your garden. How many of the plants you grow actually flower, and how many just have nice green leaves and not much else?

Those maintenance-free spiky plants and grass verges much preferred by commercial landscapers need to be replaced by (or at least intermingled with) flowering plants of all shapes and sizes. Mix it up between the exotics and the natives, the idea being to produce a cornucopia of forage that will suit all sorts of beneficial insects. Now don't get scared, we are not talking rocket science (or should I say botanical science?) when it comes to making changes—these are simple.

The things you plant in your backyard, on your balcony or by the roadside can be beautiful, practical, low-maintenance and also provide excellent forage and habitat for a whole range of insects and birds, and even human food—all at the same time.

You don't need much. A couple of old cooking oil tins make great planters for herbs, so go foraging at the rear of your nearest fast food restaurant and see if they're discarding old tins. You'll need to cut the top off and punch some holes in the bottom. Some people also find nice containers in op shops, or use old teapots or even old boots; almost any container can be used, even those re-usable shopping bags work. If that's a bit too adventurous for you, just buy some pots at your nursery or hardware store.

When you're choosing pots, plastic ones are superior to anything porous as they hold water better and won't wick it away from the roots. Just make sure they can drain and avoid the expensive self-watering pots, as these are often just a gimmick. The easiest things to grow are herbs. They'll do well in small pots and just need a bit of sun, so try things like basil, borage, rosemary, rocket—anything you would like to use in cooking. (I've listed a whole lot of bee-friendly herbs in the next chapter.) Just make sure you let some of them flower, as that's the reason you're growing them in the first place.

Just about anything that flowers will be good for all sorts of pollinators, and your aim is to have at least two plants flowering in your garden all year round—not just in spring—to provide a continuous food supply. In the following chapters we give you lists of specific plants for all types of garden spaces. They don't need to be native species, they can be anything you like the look of, just as long as it flowers. Even some weeds are fine, although you don't want to be propagating a harmful species, so check with the council in your area before you plant a whole backyard of dandelions (*Taraxacum officinale*), which are classified as a noxious weed in some places.

Once things start flowering, resist the urge to spray the caterpillars and aphids with poison—use

The things you plant in your backyard, on your balcony or by the roadside can be beautiful, practical, low maintenance and also provide excellent forage and habitat for a whole range of insects and birds, and even human food.

Look for planting
opportunities outside
your front yard or
apartment. Roadside
verges and nature
strips can provide fresh
frontiers for forage and
habitat.

Water features are a
sure way of attracting
birds and insects into
your outdoor area. This
apartment terrace offers
a convenient watering
stop at tree height.

Left: Lily pads provide a safe landing pad for bees to collect water. Right: Salvia is a buffet favourite for bees.

the natural methods outlined in the chapter on pest control. Don't forget that nature equipped you with some of the best pest-control devices there are—at the end of your arms—and although the idea of picking off caterpillars or snails might be a bit confronting to some, whack on a pair of gloves and give it a go. It's surprisingly satisfying, and the bugs can be fed to chooks or added to compost so they don't go to waste.

WATER FOR YOUR POLLINATORS

Apart from being just beautiful to watch and listen to, birds are great insect-control devices. So you'll want to be attracting them, if you can. Remember, they need water (as do bees)—so get a water source for both, perhaps a bird bath, and make sure it's not too deep, with sloping sides to allow different bird species to bathe and splash around. Bees will also enjoy the water, and take it back to the hive for their cooling system.

Make sure there are some floating platforms like corks or a sloping bank for bees to land on while filling up. (I've even heard of bees visiting a chicken waterer for a drink, which was great for the bees but

less so for the chickens, who were a bit wary of the buzzing invaders.)

YOUR GARDEN— POLLEN BUFFET AND HUMAN HAVEN

When you're thinking about what to plant, remember that your plot needs to meet both your bugs' needs and your own. You want to plant a pollinator cornucopia that provides a year-round, all-you-can-eat smorgasbord for your bees. Your garden should indeed contain a mix of flower-head shapes, colours and flowering patterns so you cater for as many insects as possible, but you don't need to be a martyr to your insect friends. Not all plants suit everybody, so the idea with these next few chapters is to pick and choose the plants you like. For example, if you're not a fan of lavender (and there are, unbelievably, people who aren't) then don't plant it. Or if you don't like rosemary because it smells too medicinal, then leave it out. The bee police are not going to visit and insist you put some in.

Another consideration for laying out your green space is allocating a spot for viewing your garden visitors. After all, the aim of this garden is to attract

all sorts of insects, so if you've gone to the trouble of luring them in, you might as well enjoy their merry to-ing and fro-ing. Set aside some space for a chair or two so you can relax and watch Mother Nature in action—providing the entertainment is the return they give for a meal.

Taking it one step further ... I like to think that the garden should be like a stage, laid out around your allotted viewing area, with you occupying the best seat in the house. Whether your garden is a little primary school stage or the opera house will depend on the space you have, and your imagination.

INSECTS LIKE SMALL GARDENS TOO

My own garden is in an inner-city suburb. It's conveniently located outside my kitchen on the roof of the garage—think square, brown 1970s tiles nicely offset by the speckled brick of the surrounding walls. My planting containers are a mixture of custom-built garden beds, and traditional ceramic pots, as well as pots made of old cooking oil drums, metal garbage bins and stainless-steel honey extractors. And you should see the swarms of bees, wasps, dragonflies and all manner of insects and birds that buzz around at all hours of the day. Bigger is not better, it's true. And landscaping flair, while comforting to the human eye, is fairly irrelevant to the bee, at the end of her working day.

My planting containers are a mixture of custom-built garden beds, and traditional ceramic pots, as well as pots made of old cooking oil drums, metal garbage bins and stainless-steel honey extractors.

Whether your space is a small courtyard, a side passage or even a balcony, there are lots of plants you can bring home and make your own. Bear in mind, of course, that many flowering plants won't flower or do well without at least a few hours of sunlight a day.

Though there are plants that thrive in bigger gardens, there are plenty that do better in small backyards or balconies, because of the microclimate created there. These compact spaces are often surrounded by buildings (that retain heat and block wind), which also help prevent frost.

If your space is super-tiny, like a corner of a Juliet balcony, don't be discouraged. A couple of flowering pots will still make a difference, especially if your neighbours have some as well. All those pots on adjacent balconies add up to be, in effect, one large, juicy garden.

If you're working with a small space, then pots are a great option, as you can grow an awful lot of plants and keep moving them around so the ones in flower are at the front of the pollen tuckshop, and the vegetative ones are at the rear.

At my local farmers' market the plant lady sells woven plastic bags of herbs all planted together that she sells as gift sets. They're about 50 cm in diameter and look fantastic, packed with all sorts of herbs—a really practical way of adding some plants to a balcony.

LARGE GARDENS

If you're lucky enough to have a garden larger than a mere balcony or a deck, and actually have proper soil, then there is so much you can do to support bees and other beneficial insects in your area. The opportunity with a generous space is to think big—maybe even think reuse. For instance, do you have a tennis court but no longer play tennis? With a little imagination you could fill it with a fantastic kitchen garden. I used to have some beehives

Don't worry too much about spending a fortune on pots. I drill holes in the bottom of old empty oil tins—the 'olde-worlde' artwork adds a certain charm.

Don't forget to set up a spot (or many) where you can sit and watch the bees and other garden goings on.

installed in the backyard of a large house in a well-to-do suburb, where the tennis court had been transformed into a productive garden complete with chook yard. There was little evidence of its past use apart from a rectangular shape and excellent night-time lighting.

The same applies to an under-utilised swimming pool. Do you have one that's currently hunkering outside, breeding mosquitoes or costing a fortune in electricity and chemicals? You can give it a second life by turning it into a productive part of the backyard that gives back rather than takes from your pocket. Grow your own produce and help local wildlife while you're at it.

Earlier, I mentioned setting your garden out like a theatre with you having the best seat in the house for all the insect action. You can really go to town with this idea in a large space, except instead of one performance space you could have a number of them—grouping the plants both big and small so your relaxation spot gives you multiple views to enjoy.

All the small garden plants can of course be used in your large garden—they will give different heights to your garden and make it more interesting to the eye.

If you're planting an area that doesn't get a lot of traffic, consider putting a path through the middle and filling the rest of your space with non-grassy plants or ground covers. An excellent choice is thyme, which has a number of ground cover species commonly called wild thyme or creeping thyme. Thyme smells divine and bees love the flowers for pollen and nectar.

The next few chapters provide a guide for bee-friendly plants suitable for all types of gardens—from small balconies to large plots—and different types of plantings to suit your needs or taste. Mix them up and create a rich variety—the bees will love you for it.

Bee goggle garden

There are a few basic things that need to be taken into account when planning your garden, so whack on those bee goggles again for a minute while you get your note pad out.

Where possible, grow your plants together in clumps and layers, as bees prefer an organised view with the same sort of plant grouped together. Use a mix of local native plants, as they're ideally suited to your environment, along with heirloom varieties of exotic plants, which are likely to be more bee-friendly. When choosing the flowering plants, a mix of flower shapes is important, as some bees have a preference for one sort of flower over another.

Where possible, establish your garden in a sunny spot, as bees generally prefer flowers that are in full sun, and select plants that flower at different times so there is something budding almost all year round. This also gives bees a varied diet, which is important for good nutrition.

This honeybee is making a beeline for some Thai basil. As well as being a great addition to Asian dishes for us, its a big favourite with bees.

Chapter Seven

Strawberry

EDIBLE PLANTS

Let's start with herbs. Many of us are keen herb gardeners and the good news is that herbs are not only easy to grow (and they do so happily in pots), but are an excellent source of forage for bees.

But the problem is we've all been taught to remove the flowers as soon as they appear on our herbs, to keep the plant producing leaves (which is what we use for cooking). Basil is an excellent example of this. Unfortunately, removing the flowers means that our bees and other insects have nothing to forage on, and while your pasta sauce will taste great, you may as well be growing rocks for all the good it will be doing for local bees. A great compromise with bushy herbs like basil is to leave the flowers on half the bush and remove them on other half. That way you get your leaves and the bees get their flowers—a win-win.

No matter what size garden you have, it's easy to have a few pots of your favourite herbs growing so you can snip off a bit for your next meal. Some things are ridiculously easy to grow and you'll wonder why you ever paid for it before. Growing herbs is great for the environment as well, as there is a lot of fuel and water invested in the simplest of herbs from the time it leaves a commercial herb farm to the time it lands in your kitchen.

If you have a larger space and like the idea of being more self-sufficient, then growing some of your own fruit and veggies can fill your fridge and sometimes the fridges of your neighbours with an ongoing bounty of home-grown produce. Nothing can compare with food you grow yourself; I once held a dinner party where every course (apart from the basics like flour and butter) consisted of food I had personally grown and harvested. It was a rewarding experience like no other.

You could consider growing a pumpkin, zucchini or cucumber vine—bees will pollinate these and find them useful for pollen—but beware: you might be looking for Jack to climb this vine instead of a bean stalk, as they do have a tendency to become voracious consumers of garden real estate. All of these can suffer from powdery mildew, so avoid watering the leaves.

If you do have a veggie patch in mind, here are a few tips to help keep the plants in tiptop condition and your garden productive as a result.

CROP ROTATION

This is a must do. Many plants, including potatoes and tomatoes, can be damaged by things living in the soil, like nematodes (microscopic parasitic worms), which affect the root of the plant, causing poor performance and death. If you practise crop rotation (where you don't use the same soil every year for the same plant), you can help reduce the incidence of parasites in the soil, as they will miss out on their favourite meal for a season.

You can also try planting 'green manure' crops. These are fast-growing grass-like crops, planted annually with the purpose of smothering weeds and increasing the organic matter and other essential elements like nitrogen in the soil. Barley, oats or sorghum and a legume together will do the trick. These crops are also excellent for bees as

Clockwise from top: Containers are a great option for growing vegetables and herbs if you live in an apartment; parsley flowers; chive flower.

Purple coneflowers are companion plant powerhouses. They produces big amounts of nectar and pollen—making them a real bee magnet—and seeds that are irresistible to small birds.

Left: Edible greens happily grow in tubs. Right: Thyme flowers are a sure way of attracting pollinators.

they flower and produce good nectar and pollen. A green manure is ultimately mulched and dug back into the soil, providing organic matter as well.

COMPANION PLANTING

This can be broken down into a couple of types and is a very effective way of helping your veggie (or decorative garden) repel pests and flourish.

Some plants help attract pollinators or predators to your garden, encouraging a healthy diversity of critters. One easy way of planting insect attractors is to grow herbs. Other plants are used as decoys or masks. Masking plants disguise the smell of your preferred plants and confuse the insect pests that might otherwise attack them. A good example is planting garlic near roses to deter thrips, aphids and other pests. Or placing basil and tomatoes side by side is said to deter pests and

improve flavour—but it could be that the basil is attracting more bees and improving the flavour of the tomatoes through effective pollination.

Decoy plants are highly attractive to pests. Though I am fond of Dad jokes and am often heard to ask, 'Are you casting nasturtiums?' if I was a plant, I would be saying, 'Please throw heaps at me because bugs can't resist them'.

Some plants help attract pollinators or predators to your garden, encouraging a healthy diversity of critters.

You don't need a huge space to create your own edible plant garden. Pots and small courtyard garden beds will also work well.

GREAT HERBS FOR BEES

With enough sun, most herbs are really easy to grow—they will grow like weeds in the right circumstances and you'll find new ones popping up to replace the ones that are reaching the end of their cycle and becoming too woody. You have to resist the temptation to trim off all the flowers and leave at least half behind so you get the seeds to allow the plant to continue, and of course provide pollen and nectar for the beneficial insects.

Lavender
Lavandula spp.

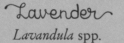

The best lavender species for most of Australia is French lavender (*Lavandula stoechas*). This one is more resistant to humidity and less susceptible to fungal problems than other species, such as English lavender (*Lavandula angustifolia*), and once established will be mildly drought tolerant. Lavenders, generally, are great for bees because they can have very long flowering periods. They're great for us, too, because they look pretty and can be harvested and dried to make pot pourri so your undies smell like granny's house.

Bees that sup on lavender produce good lavender-fragrant honey, but you would need fields of it to actually taste it in any honey your bees make.

Sun: Full ✱ **Soil:** Slightly alkaline, well drained ✱ **Planting:** Pots, containers, garden ✱ **Size:** 50 cm or more tall ✱ **Flowers:** Summer

Sweet Basil
Ocimum basilicum

This is one of those plants that you're taught to pinch the flowers off so it doesn't bolt (go to flower), but don't! Let it grow and flower; let it bolt if it needs to. When it flowers and drops seeds these will sprout into new plants, making for an everlasting supply. Basil is an amazing bee attractor, with a sweet aniseed-like smell that bees of all sorts seem to find irresistible, once your plant gets going.

Basil needs a generous sized pot to give it room to grow up big and bushy (it can become a bit of a tree if it's given free rein), bearing in mind the bigger the bush, the better the bee magnet. It's a prolific flowerer and of course there are all those leaves for your pasta or pesto. And it will grow almost anywhere there is sufficient sun (five to six hours a day).

Basil originally came from India, so it's remarkable that we think of Italian food first when we think of this herb.

Sun: Full to part shade ✱ **Soil:** Rich, well drained **Planting:** Pots, containers, garden ✱ **Size:** 50 cm or more tall ✱ **Flowers:** Summer–autumn

Rosemary
Rosmarinus officinalis

Another classic bee plant, rosemary has many varieties—you just need to choose the one that's right for your climate and your particular space. Some grow into big bushy plants and others are prostrate ground covers. Visit a good local nursery and get some advice—they should stock many varieties. Nearly all of them can tolerate drought or dry soil, and some of them actually thrive on it. They do prefer full sun, so don't keep them in the shadows. And as many of us who've cooked a Sunday roast would know, rosemary leaves are great sprinkled on roast potatoes. As for the bees, rosemary produces carbs for their diet (nectar) but little pollen (protein).

Sun: Full * **Soil:** Very well drained * **Planting:** Pots, containers, garden * **Size:** Depends on variety, some prostrate some bushy, up to 40 cm * **Flowers:** Winter–spring

Borage
Borago officinalis

People always talk about borage in reverential tones as the ultimate bee plant, and it's no wonder. This plant produces lots of nectar-rich, star-shaped purple flowers that bees just can't go past, as well as good pollen. The flowers reportedly restock their nectaries more regularly to increase the bee visits. (Amazingly, research shows that bees have an idea of when this happens, so they know how often to come back.) The leaves of this plant can be eaten like spinach (don't worry—once you cook it, those hairs disappear). The flowers are also edible.

Borage can be grown in partial shade and will do well in that environment. In my balcony garden I have a lot of borage that's usually buzzing with blue-banded bees. The plant has also reseeded a number of times and is taking up a larger area than I anticipated, so pots or containers might be a good option instead of a garden bed.

It's a plant that according to old lore gives you courage, even for the most daunting of tasks. Apparently, ladies, if you slip borage into a man's meal it will give him the courage to propose, so use it wisely. Clearly, borage is an essential addition to your bee garden.

Sun: Full to part shade * **Soil:** Well drained * **Planting:** Pots, containers, garden * **Size:** About 40 cm tall and 100 cm wide * **Flowers:** Spring–summer

Lemon Balm
Melissa officinalis

If you have a lemon balm in flower the bees won't leave your garden alone: they love it. With leaves that look very similar to mint (and a proclivity for spreading like mint, too) it can be grown in pots or garden beds in full sun or partial shade. Lemon balm produces both pollen and nectar. It's a decent lemon substitute in a herbal tea or in some desserts—crush a leaf and get a whiff of its lemony scent. In the past, beekeepers would rub a handful of lemon balm inside the hive after hiving a new swarm in order to help the swarm settle and to encourage them not to leave the hive. The same is done with lemongrass; the idea being that the lemony scent is a bit like queen pheromone.

Sun: Full to part shade * **Soil:** Well drained * **Planting:** Pots and containers help control its spreading tendencies * **Size:** About 50 cm tall **Flowers:** Summer–autumn

Thyme
Thymus spp.

There are many varieties of thyme, so pick one that suits your area and space. There are prostrate thymes, which spread as ground covers, and more upright versions that grow up to be regular bushes.

A traditional honey plant in Greece, where it grows in huge numbers, thyme can tolerate full sun and low water, preferring well drained soil. It's a perennial plant and a significant honey and pollen producer, which makes it ideal for your bee haven.

Sun: Full to part shade * **Soil:** Alkaline, well drained * **Planting:** Pots, containers, garden * **Size:** Many different sizes, but predominantly a small bush about 15 cm tall * **Flowers:** Spring–summer

Sage
Salvia officinalis

This is a lovely herb to have in the garden and goes well with a lot of dishes—check sage in the index of your favourite recipe books if you're not convinced. It's a hardy full-sun lover, and its blue flowers are good nectar producers. Personally, I don't like the smell of sage in my garden, but don't let that put you off, as it's a great bee plant and a few leaves in burnt butter sauce are divine. Sage is happier in a garden bed, but will also be fine in a larger pot.

Sun: Full to part shade * **Soil:** Sandy, well drained * **Planting:** Pots and containers help control its spreading tendencies * **Size:** About 40 cm tall, depending on variety * **Flowers:** Spring–early summer

Mint
Mentha spp.

Common mint can be a crazy weed so
is best in pots, where you can control its
tendency to spread. Bees seem to really love
its feathery sprays of white or pink flowers.
As for the leaves, pick some for your mint
tea, your mojito, or even just a refreshing jug
of water, with some lemon slices. Spearmint
and peppermint are part of the same genus,
and are also great bee plants.

Sun: Full to part shade ✴ **Soil:** Moderately rich,
moist, well drained ✴ **Planting:** Containers
and pots help control its spreading tendencies
✴ **Size:** Up to 40 cm tall depending on variety ✴
Flowers: Summer–autumn

Oregano

Origanum vulgare

Another great garden herb to have growing is oregano. If you let it grow freely, it produces tiny, delicate flowers in pink or white, which make it another bona fide bee magnet. Oregano honey is common in Turkey and the Mediterranean—no surprise then that it likes full sun. Oregano is just one species of *Origanum*, and is often called wild marjoram. Marjoram (*Origanum majorana*) itself (sweet and pot marjoram) has a sweeter spicier flavour, but all *Origanum* are pretty much interchangeable in the kitchen.

Sun: Full * **Soil:** Well drained * **Planting:** Pots, containers, garden * **Size:** 20–30 cm * **Flowers:** Spring–autumn

Chives
Allium schoenoprasum

These relatives of the onion family are, not surprisingly, related to garlic, leeks and spring onions. I love a bit of chives chopped into my scrambled eggs, so I always have a container growing in my garden. The large purple flowers are just gorgeous and edible as well, not to mention very popular with your local bees. Chives have a subtle garlic and onion flavour and were once hung around the house to ward off evil, so who knows, they might do the same in your backyard.

Sun: Full; can tolerate part shade * **Soil:** Rich, well drained * **Planting:** Large pots and containers, garden * **Size:** About 20 cm * **Flowers:** Mid-summer–autumn

Coriander
Coriandrum sativum

Not only fantastic in Asian dishes (for those of us who like the taste—not everyone does) it's highly attractive to all sorts of beneficial insects. Coriander flowers easily if you let it bolt, which mine always seems to do almost straight away. I've since found out it's from the stress of replanting seedlings, and if grown from seed it's less likely to bolt.

Interestingly, experiments done in the former Yugoslavia showed that yields of coriander plants increased from 800–1200 kg per hectare, up to 1800–2000 kg per hectare when beehives were placed around the crops, with the bees preferring coriander over all the other available flowers.

Sun: Full, can tolerate part shade * **Soil:** Rich, well drained * **Planting:** Pots, containers, garden * **Size:** About 50 cm (mine never gets there as I eat it first) * **Flowers:** Summer–autumn

Parsley
Petroselinum crispum

Now here is a strange one for you. Have you ever seen a parsley flower? It belongs to a special class of plant called 'biennials', which only flower in the second season when its root is substantial enough. So it will take some time before the bees get to enjoy this one, and in the meantime you'll have some lovely herby flavour to add to your cooking. In order to allow it to grow big enough to flower, harvest the outside stalks, leaving the inner ones to grow.

Sun: Full to part shade * **Soil:** Rich, moist, well drained * **Planting:** Pots, containers, garden * **Size:** 40 cm as it gets old * **Flowers:** Summer

Scarlet runner bean
flowers need to be
pollinated in order to
seed. While the flowers
are self-fertile, bees
offer the best result for
transferring pollen.

SALAD GREENS AND OTHER VEGGIES

Just because they're green, doesn't mean they won't be good for bees. Salad plants like cabbage, lettuce, rocket, bok choy, mustard greens and radishes all flower prolifically if you let them, and will be a welcome addition to the flower selection in your garden (as well as providing healthy leaves for your dinner). Here are a few suggestions to get your human and bee veggie patch started.

Rocket
Eruca sativa

I always have rocket growing in my garden; it self-seeds readily and the bees love it when it goes to flower. It's so handy to have some on hand for salads and sandwiches. I also use it like spinach and steam it as well. You need to thin it out a bit, as when it flowers and drops seed you can end up with too much growing … it almost becomes a weed, but a tasty one at that.

Sun: Full to part shade ✳ **Soil:** Well drained ✳ **Planting:** Pots, containers, garden ✳ **Size:** Up to 30 cm ✳ **Flowers:** All year round

Fennel
Foeniculum vulgare

This plant, often found by the roadside, is considered an invasive weed in Australia, so care should be taken that the seeds do not disperse and spread. It's a fantastic culinary plant with a mild aniseed flavour, similar to star anise, and bees seem to like it too. The bulb can be eaten raw or cooked, and the leaves and seeds are also used in cooking. Fennel comes from the Mediterranean, but has become common in many parts of the world, especially in dry soils near the coast and on riverbanks.

Sun: Full ✳ **Soil:** Rocky, dry ✳ **Planting:** Large pots and containers, garden ✳ **Size:** 60 cm, seeds spread rapidly ✳ **Flowers:** Late summer–autumn

Tomato

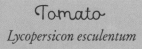

Lycopersicon esculentum

You can't have an edible garden without tomatoes. These members of the deadly nightshade family are a fruit that would have to be the most recognisable, if not most confusing because they're recognised as a vegetable. Members of this plant family require the special form of pollination, known as buzz pollination, and really benefit from visits by blue-banded bees, which can offer this service (regular European honeybees cannot). I prefer to grow the smaller grape varieties, and they often reseed and pop up again the following year. Tomatoes can suffer from fungal problems, and they benefit from companion planting with basil—together you have the basics for a fantastic pasta sauce. Heirloom varieties often taste better, so look for these when you're purchasing seeds or seedlings.

Sun: Full to part shade * **Soil:** Well drained * **Planting:** Large containers, garden beds * **Size:** Various sizes; often need staking to keep upright * **Flowers:** Summer

Beans

Fabaceae

The term 'bean' relates to whole families of plants used for human and stock feed. All beans are nitrogen fixers, meaning they take nitrogen from the air and add it to the soil—making them a natural fertiliser. Most beans require support to grow. I use mesh to support my vines.

Historically, beans used to be grown alongside corn, and the beans would wind up on the corn stalks with the beans providing nitrogen for the corn. Squash was also grown alongside, using the corn and beans as a wind break and to help retain moisture—acting as a live mulch of sorts.

Beans are called heliotropic plants because their leaves follow the sun during the day to maximise the exposure. Not all beans need pollination; some are self-fertile. So, if you're planting them for your bees more than your table, make sure the variety you are planting is suitable.

Sun: Full * **Soil:** Well drained * **Planting:** Large pots, garden beds * **Size:** Various sizes; often need staking to keep upright * **Flowers:** Depends on variety; summer

Planting a mixture of herbs and flowers (zinnias in this instance) in your veggie patch will encourage good bugs and birds to move in and take care of pollination and pest control.

BACKYARD FRUIT FARMS AND ORCHARDS

I can remember an ever lasting supply of almonds and figs from our backyard when I was a child, and scoring my initials in developing passionfruit to claim it as mine. This usually led to backyard arguments, but it was great fun.

You can grow just about any stone fruit, citrus or pome fruit tree in your backyard to supply nectar and pollen (and fruit as a bonus). Check the variety you're purchasing to make sure it's not going to be a huge monster of a tree with too much fruit for you ever to use, as you don't want waste fruit and create a fruit-fly problem.

Plenty of fruit and nut trees can be grown in large containers and most varieties produce nectar-laden flowers that bees love. You need to keep on top of pruning, though, so they don't take over.

Blackcurrant
Ribes nigrum

A cool-climate berry, blackcurrants are very high in vitamin C and very productive with a single bush capable of producing 5 kg of fruit. Blackcurrants are small bushes that don't need staking for support, but like the raspberry, all old canes should be pruned to allow new wood the next year and plenty of fruit.

Sun: Full, cooler regions only ✽ **Soil:** Well drained **Planting:** Large pots and containers, garden ✽ **Size:** Small scrubby bush about 1.5 m tall ✽ **Flowers:** Spring

Strawberry
Fragaria × ananassa

Who doesn't like fresh strawberries? They're irresistible to adults, children and birds, so plan to protect them from whichever you think will be the greatest threat.

Some strawberry varieties fruit twice a year, in the spring and again in autumn, which sounds like a bonus to me. Like a lot of the berries, it's essential you seek out certified virus-free stock to get the best results, and keep the runners under control, as strawberries don't like to be crowded. In the warmer parts of the country, shade cloths might be necessary to prevent sunburn, but make sure there is plenty of ventilation, as fungus is the enemy.

Old-fashioned strawberry pots are still used by some people, but also consider a hanging pot with the fruit hanging over the side—it's a much easier way to grow them. Whatever pot you use make sure you water uniformly, carefully avoiding the foliage. Plant in winter and harvest in spring.

Sun: Full; shade in hot sun regions ✽ **Soil:** Well drained with plenty of vegetable matter ✽ **Planting:** Pots, containers, garden ✽ **Size:** 10–20 cm ✽ **Flowers:** Spring–summer, depending on variety

Blueberry
Vaccinium corymbosum

These berries do best with buzz pollination, so blue-banded bees will make a difference to the fruit yield, which can be about 7 kg per bush. There are warm-climate varieties, so seek out advice at your nursery and purchase the appropriate one for your region. The berries go through a number of colour changes before reaching the dark blue colour that also indicates they're ripe.

Sun: Full * **Soil:** Well drained * **Planting:** Large pots and containers, garden * **Size:** 1–2 m * **Flowers:** Late spring–autumn

Raspberry
Rubus spp.

Bees love raspberry bushes and, depending on the cultivar, they can flower from spring to mid-summer for one to two months. They're a significant nectar producer that bees can't leave alone.

They're best suited to cooler regions, so choose a type that suits your garden—some need staking for support as they become laden with berries. They're suckering plants and can get out of control if let run wild, so either plant in a large container or keep an eye on them.

When purchasing plants, look for certified stock to ensure they're disease free. Pruning is essential to get a good crop each year (as only new growth will produce fruit), so once fruiting has finished, remove all the canes that produced raspberries through the season.

Sun: Avoid hot midday sun; cooler regions only * **Soil:** Well drained * **Planting:** Large pots and containers, garden * **Size:** About 2 m tall * **Flowers:** Spring–mid summer

Lemon, lime and orange
Citrus spp.

For many years so many households had a lemon tree in their backyard. It's time to bring them back.

Citrus trees can be a great addition to a small–medium space as they can thrive in (big) pots, are great flowerers when they get going, and bees just love their blossoms. When your lime crop is ready, make sure to cut yourself a slice of lime to drop in your G&T and sip while watching the bees come and go from your garden. Seek out dwarf varieties for containers and smaller garden areas.

Sun: Full * **Soil:** Well drained * **Planting:** Large pots, garden beds * **Size:** 1.5–5 m * **Flowers:** Depends on variety

Passionfruit
Passiflora edulis

I once had somebody thank me for installing some beehives because their passionfruit had finally produced fruit after four years of lovely but fruitless flowering. These Australian backyard staples have extensive root systems and really need to be grown in the ground rather than a container, with a strong trellis for support. There is some old gardening lore about planting a passionfruit over some offal— this is probably just to give the vine extra iron, and that can be achieved with an iron-rich fertiliser.

Passionfruit are subtropical plants, so will grow in most parts of Australia, although they do need a nice warm sunny spot if you're in a colder part of the country, and shelter from the hot sun if you're in a hotter area. Some varieties are self-pollinating, but still produce lovely flowers that honeybees will appreciate.

Sun: Full in colder areas; part shade in hotter areas * **Soil:** Well drained, sandy soil * **Planting:** Garden * **Size:** Depends on pruning and size of trellis; can tend to woody and dense if not pruned annually * **Flowers:** Late winter–spring, late summer–autumn

Almond
Prunus dulcis

These ancient nut trees were cultivated in the Mediterranean as long ago as 4000 BC and have been popular ever since. These days, California is the world's largest producer of almonds, which were introduced into the USA in the mid-1800s. You need two trees to obtain fruit, but both trees can be planted together so they intertwine to become one large tree. Almonds need to be planted in the ground, not in pots, and can become very large. Careful attention to location and pruning is needed to keep them under control.

Almond trees flourish in mild, wet winters and hot, dry summers in full sun, and are unlikely to fruit if your region doesn't have that climate. They also need irrigation, as they're voracious water consumers and require good drip irrigation (their water consumption is a constant source of criticism for the commercial almond industry). Bees, however, love the flowers and will collect both nectar and pollen from them.

Sun: Full ✳ **Soil:** Well drained, but plenty of water in the form of drip irrigation ✳ **Planting:** Garden ✳ **Size:** Can be a large tree or pruned back ✳ **Flowers:** Winter

Avocado
Persea americana

These trees not only produce the fleshy, creamy green fruit that we pay a fortune for in the supermarket, but they also flower profusely and are great nectar and pollen producers. They come from South America and a number of varieties are ideal for growing anywhere without frosts. Just check the size of the species you're growing, as some can be very large trees indeed, reaching well over 15 metres. Until recently, my neighbour had two massive avocado trees in his backyard, and I'm sure their removal is part of the reason for the lack of honey in my backyard beehive.

Sun: Full ✳ **Soil:** Well drained ✳ **Planting:** Large pots and containers, garden ✳ **Size:** Small tree ✳ **Flowers:** Spring

Apples
Malus × domestica

Apples are a great addition to a backyard, providing crisp, juicy fruit that you have to taste to believe (especially if you're used to buying apples from a supermarket). Apple trees are usually grafted, so spend some time looking for a variety that will suit the space you have in your garden and the type of apple you prefer to eat.

Apples will need a tree growing nearby to provide pollen (its polliniser) and often a crab apple is planted to fulfil this role. It's possible to have a tree with multiple varieties grafted onto one rootstock (even red and green on the same tree), and this can also include the polliniser. Don't waste the fruit of any of these trees—if you have an overly abundant bounty, swap it with your neighbours to create your own little farmers' market.

Sun: Full * **Soil:** Rich, well drained * **Planting:** Large pots and containers, garden * **Size:** 1.5–8 m * **Flowers:** Spring

Banana
Musa acuminata and *Musa balbisiana*

I once successfully grew bananas in a small pot on my balcony—it's the first crop I ever produced. They actually don't need pollination to fruit or reproduce, but nonetheless produce abundant flowers when they're fruiting, and bees love the pollen and nectar from them. The supermarket bananas we eat have been bred to be seedless and are actually infertile; they're replicated by splitting off rhizomes and repotting them.

An interesting fact is that nearly all cultivated bananas are clones of a plant grown in England in 1830, the Cavendish banana, which makes them very susceptible to disease. For decades the commercial banana of choice was the 'Gros Michel', but in the 1950s it was practically wiped out by the fungus known as Panama disease, or banana wilt, and the Cavendish, which was not susceptible, was grown to replace it. It's very possible that a different problem may wipe out the Cavendish banana industry in the future, so banana trees are regulated and you need to purchase stock that's certified. Bananas grow best in a warm, frost-free, coastal climate. They need all-day sunshine and moisture.

Sun: Full sun, low wind * **Soil:** Well drained with lots of vegetable matter * **Planting:** Large pots or containers, garden * **Size:** Tends to sucker and can be a large messy plant * **Flowers:** Late spring–autumn

Australian gum trees are prolific flowers and provide a rich source of forage for native bees. This red flowering gum is also a favourite with nectar-feeding parrots.

Chapter Eight

Banksia

NATIVE PLANTS

Native plants are crucial to your bee-friendly garden — it's really important that your garden consists of a mix of exotic and native plants to keep food sources available all year around. Some of our lovely native bees have evolved to depend on specific native plants for forage and would perish without them.

SMALLER PLANTS

A great thing about adding native plants to your garden is that you can chose plants that exactly suit the region you live in—from the amount of sun to the type of soil. Here is a selection of smaller plants that grow in a broad climate range and are particular bee favourites.

Native daisy
Brachyscome spp.

There are some wonderful species of native daisies that make for pretty pot plants. They flower prolifically and the flowering period lasts a long time, offering pollen and nectar to bees and other pollinators. Some species flower twice a year. It's a large genus of 70 to 80 plants that can be annual and perennial ground cover plants as well as small shrubs, so do some research and work out which ones best suit your garden area. Some varieties flower in spring and again in summer.

Sun: Full to part-shade * **Soil:** Dry to moist * **Planting:** Pots, containers, garden * **Size:** From ground covers to bushy plants, depending on variety * **Flowers:** Generally spring–early summer

Native sarsaparilla
Hardenbergia violacea

These gorgeous Australian natives appear as both climbers and small bushes, depending on the variety. They're mainly available with white, purple or mauve and pink flowers. They're legumes with pea-like flower sprays and are also nitrogen fixers. It's best to pick a variety that will do well in your area, as not all are suited to all regions. As they flower in winter, they can give bees a welcome food top-up in an otherwise lean time.

A great variety is 'Happy Wanderer', an evergreen climber that flowers in late winter to spring, making it an excellent choice to add both some colour to your garden and some forage for the bees. Prune after flowering to promote compact growth and better flowering next season.

Sun: Full for best flowering results * **Soil:** Well drained * **Planting:** Pots, containers, garden * **Size:** Depends on variety; some prostrate forms to large climbers * **Flowers:** Late autumn–spring, depending on variety

Native violet
Viola hederacea

This spreading native ground cover is great at covering rockeries or even as a lawn replacement for those spots where lawn will just not grow. It has a carpet of small green leaves and flowers almost constantly during the warmer months. It needs frequent water so may not be a good choice in dryer parts of the country.

Sun: Shady, cool * **Soil:** Moist * **Planting:** Pots, containers, garden * **Size:** Spreading ground cover 10 cm tall * **Flowers:** Spring–autumn

Native fuchsia
Correa spp.

Correas are known as native fuchsia due to their elegant, tubular flowers. These bee- (and bird-)attracting evergreen shrubs are available in a variety of colours, and flower through autumn and winter, which makes them a desirable addition to balconies and small gardens. Correas are best grown in clumps of four or more plants to give a decent group of flowers for insects to forage.

Sun: Does well in shady locations. Avoid in very dry or tropical areas * **Soil:** Reasonably well drained * **Planting:** Pots, containers, garden—makes a fantastic low hedge * **Size:** mostly bush plants to 100 cm * **Flowers:** Autumn–spring

Fairy fan-flower
Scaevola aemula

Fairy fan-flowers produce blooms for a long period of time, some for nine months of the year, making them a great option for long-term flower production. Colours range from white to blue or purple. They're perfect for all bees, but native bees really love this one—the whole plant becomes a riot of flowers when it's in bloom and makes an amazing ground cover. There are many varieties (more than 70), from very low ground covers to taller shrubs.

Sun: Full * **Soil:** Well drained * **Planting:** Pots, containers, garden—spreading * **Size:** 30–100 cm * **Flowers:** Late winter–autumn

Grevillea
Grevillea spp.

This is a huge group of more than 300 species according to the experts, available in numerous shapes and sizes, so that there is a grevillea for almost any conceivable garden situation. Often with short fern-like foliage and, in most cases, very nectar-rich flowers, they tend to be bird and bee magnets. Many flower from winter to early spring, while others will flower and provide nectar all year round. You can actually shake a grevillea flower onto your hand to see the nectar drip out. They can be grown in large pots and containers but do best in the ground.

Sun: Full; can tolerate some shade * **Soil:** Well drained * **Planting:** Pots, containers, garden * **Size:** Low ground cover to large bushy trees * **Flowers:** Winter–early spring

Christmas bush

There are plants called Christmas bushes in every state of Australia, almost all named because of the red flowers that appear close to Christmas, including *Prostanthera lasianthos* from Victoria and *Bursaria spinosa* in South Australia. A particularly popular Christmas bush is *Ceratopetalum gummiferum* (shown here) from New South Wales, which produces flowers that change from white to rusty red after flowering, heralding the arrival of Christmas. It's a favourite plant with all insects, in particular native bees.

Sun: Full for best flowers * **Soil:** Well drained * **Planting:** Containers, garden beds * **Size:** 5 m * **Flowers:** Spring–summer

Pincushion tree
Hakea spp.

There are some 140 shrubs and small trees in the hakea genus. Some can be confused with grevilleas (their cousins) as the flowers can be similar, but others, like the pin-cushion hakea (*Hakea laurina*, shown here) look completely different. Some grow to 15 m in height and others are small scrubby bushes. All are bird and bee magnets. Some, like the two-leaved *Hakea trifurcata*, commonly called kerosene bush, are exceptional nectar producers.

Sun: Full * **Soil:** Very well drained (most are from Western Australia and like sandy soil) * **Planting:** Garden * **Size:** 3–4 m to small trees * **Flowers:** Autumn–winter

Local council nurseries

Don't be afraid of adding natives to your garden, as there are many more plants available than you might imagine that will suit any type of garden. A good place to start when thinking about natives is a local council nursery, as many have plants for sale that are indigenous to your area.

TREES

Australia is an incredibly nectar-rich country, so it's easy to select almost any native tree and have good insect habitat in your backyard. The plants below are really only the tip of the iceberg, with many, many larger plants that I've deliberately left out as they're just too big for the average backyard. If you have a few acres, then look at blue gum, bloodwood, ironbark, scribbly gum, grey gum, yellow stringy bark and brush box. These are really big trees, but fantastic nectar producers.

Some of our native trees have incredibly long flowering periods, which make them a bounty for bees. Some trees, like the broad-leaved paperbark (*Melaleuca quinquenervia*), a once common verge tree in Sydney, flowers for months and months with a fresh creamy spray every time there is decent rain ... and bees love it for a good stock-up before winter.

Native hibiscus
Alyogyne huegelii

Most people don't think of hibiscus when they think of Australian natives, but we do have a local version that's a favourite with our bees and birds, and produces magnificent flowers of yellow, white, pink, purple or red. These fantastic blooms are, however, short-lived, as each flower only lasts one day. It's a good thing they're prolific flowerers.

Sun: Full ✳ **Soil:** Well drained ✳ **Planting:** Garden ✳ **Size:** From small shrub to large tree of 15 m ✳ **Flowers:** Some flower all year round

Macadamia
Macadamia integrifolia

It you haven't seen a macadamia in flower then you might think I'm nuts for suggesting they're a good tree for bees. They flower in amazing sprays of white flowers that bees, in particular the tetragonula species, just can't leave alone, which leads to plenty of delicious fruit. They're native to Australia and will grow from two to 12 m tall, so you better have a good ladder to harvest all the nuts.

Sun: Full to part shade ✳ **Soil:** Well drained, sandy to heavier clay ✳ **Planting:** Garden ✳ **Size:** Up to 12 m ✳ **Flowers:** Winter–mid-spring

Lemon-scented tea-tree
Leptospermum petersonii

This is a charming Australian native tree with a tendency to shrubbiness and lovely lemony-scented leaves. It's grown in plantations for its essential oil and the leaves are sometimes blended to make a lemon tea. The new foliage is copper-bronze in colour, with many small white flowers that decorate the tree in early summer. Like many of the leptospermums, it's a terrific bee tree. It's on the larger size (growing to 5 m), but can be pruned back to maintain it as a smaller tree or hedge. Another good variety that bees love is *Leptospermum polygalifolium*, although any of the leptospermum species seem very popular with pollinators.

Sun: Full to part shade * **Soil:** Well drained * **Planting:** Garden * **Size:** 5–10 m (dwarf varieties available) * **Flowers:** Spring–early summer

Broad-leaved paperbark
Melaleuca quinquenervia

This paperbark is a common street tree in my suburb and produces an abundant nectar flow when it flowers from spring to autumn, with a new burst of flowers every time it rains. Paperbark honey is a lovely blend, and it's a great tree for bees to stock up on before winter sets in. Like a lot of our native trees this one has become established overseas, so much so that in the Everglades in Florida, it's now considered a weed.

Sun: Full to part shade * **Soil:** Well drained soil * **Planting:** Garden * **Size:** 5–20 m, depending on variety * **Flowers:** Spring–autumn

꙰Bottlebrush꙰
Callistemon spp.

The bottlebrush is an Australian native that you'll now find growing across many parts of the world. Its distinctive flowers resemble bottle brushes, hence the name. There are many varieties of bottlebrush in both big and small options, with colours ranging from the traditional red through to pink, mauve, cream and green. All of them flower profusely, and will attract birds and bees with their nectar-rich flowers. 'Eureka' is a good choice for a small dense bush of three metres. 'Rick's Red' is a great choice for a long flowerer, and it can also be used as a hedge.

Sun: Full to part shade * **Soil:** Well drained * **Planting:** Garden * **Size:** Up to 5 m, depending on variety * **Flowers:** Generally spring–summer

Water gum
Tristaniopsis laurina

This is quite a common street tree near my house, and it's generally found on the eastern coastline of Australia. When it's flowering the bees can't get enough, and the honey they produce is a strong-smelling, thick brew that's the same yellow as the flowers that cover the tree in a yellow blanket. It's really tolerant of drought despite its name, and seems to do well in city centres, which would usually have poor soil.

Sun: Full to part shade ✳ **Soil:** Well drained ✳ **Planting:** Garden ✳ **Size:** Up to 15 m ✳ **Flowers:** Summer

Some of our native trees have incredibly long flowering periods, which make them a bounty for bees.

Banksia
Banksia spp.

Another classic Australian native, the banksia is a very important nectar producer with about 170 species ranging from huge trees to tiny versions you grow in pots. Like the grevillea, there's sure to be one that fits your backyard. The hairpin banksia (*Banksia spinulosa*) is a popular one with bees. Despite the image of the big bad Banksia men from *Snugglepot and Cuddlepie*, with their abundant nectar flows, banksias are very beneficial trees to have in your insect garden.

Sun: Full to part shade ✳ **Soil:** Well drained ✳ **Planting:** Garden ✳ **Size:** Small shrubs to large trees of 25 m tall ✳ **Flowers:** Summer, autumn, winter, depending on variety

GRASS ALTERNATIVES

The average modern landscape is often designed with low-maintenance requirements in mind, but sadly these are often made up of plants with minimal flower displays, which means bees have many fewer nectar sources to attract them and keep them healthy.

In previous chapters I've had negative things to say about grass and the insect desert it produces. Luckily, there are easy alternatives to grass. Clover is one example. This is a hardy plant that produces abundant nectar and pollen, and is a delight for your local bees. It's also a nitrogen fixer, meaning it has nodules on its roots that take nitrogen from the atmosphere and restock the soil with a stable form of nitrogen—in effect, a natural fertiliser. Any surrounding companion grasses can absorb this natural nitrogen through the roots.

There are native grasses that do well instead of lawn, such as wallaby grass or weeping grass; and there are even commercial wildflower lawn alternatives, some of which are shade tolerant, so they'll grow well under tree canopies and can be walked on just like grass.

Flax Lily
Dianella revoluta

Flax lilies are important plants for a number of our native bee species. They're also dependent on native bees for pollination, as they require sonication to release the pollen. They're a native grassland species and form dense clumps when given the space to spread. In addition to *Dianella revoluta*, species include *Dianella caerulea, Dianella longifolia* and *Dianella tasmanica*. Once established they're drought tolerant and able to cope with a variety of conditions.

Sun: Full to part shade * **Soil:** Well drained * **Planting:** Pots, containers, garden * **Size:** Up to 1 m * **Flowers:** Spring–early summer

Pigface
Carpobrotus spp.

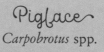

For a different look (with an unflattering name) you could try pigface. This is a creeping succulent with large fleshy leaves and large, purple, daisy-like flowers that create a wonderland for bees. It often grows on sand dunes and is extremely drought tolerant—so if you're hopeless with the watering schedule this could a plant for you. When I was a child we believed that rubbing the sap of pigface on warts would make them go away... not sure if this actually worked, but plenty of my friends tried it out. Indigenous Australians have been using its juices for aeons, mostly as a salve for stings or burns. These days you're more likely to see it on a menu somewhere, as it can be eaten raw or cooked and apparently adds a strawberry or fig kind of flavour to meat dishes.

Sun: Full * **Soil:** Well drained * **Planting:** Garden * **Size:** Ground cover, spreads readily * **Flowers:** Spring–summer

Salvias come in many shapes and sizes, and most are great for bees. They prefer a sunny spot with well drained soil, and should be deadheaded regularly in the summer months.

Chapter Nine

Peony

EXOTIC
PLANTS

As far as bees are concerned, it's all about having a mix of flowering plants in your garden. To provide that insect and bird feast you need plants that are flowering all year round, but if native plants are just not your thing you can still provide for the local insect and bird populations by planting exotics.

When choosing flowering plants it's good to keep in mind the design of the flowers, and therefore their accessibility to bees. So-called 'single flowers', with just one ring of petals and exposed reproductive parts in the centre, generally produce large amounts of nectar and pollen that bees can easily get to. Examples of single flowers are lilies and violets. Plants in the daisy (Asteraceae) family, such as coneflowers (*Echinacea* spp.), produce composite flowers that often have the appearance of single flowers and are usually good pollen producers.

'Double flowers' (flowers with more than one ring of petals) can still have plenty of food stores for bees, but are much more difficult for the bees to reach. Most garden roses are double flowers (in fact only the single-flowering *Rosa canina* and *Rosa rugosa* are good bee plants), as are carnations, camellias, peonies and geraniums—your typical 'Granny's garden' plants. Having more single flowering types and daisies, than double flowers will increase your garden's bee appeal.

Quite a few of the plants listed in this chapter have been used in gardens for a long time and may be deemed old-fashioned by some. But believe me, Grandma and Gramps were onto something when they put these in the backyard, so do like they did and do your bit for bees as well.

Left: This peony is an example of a double flower. It's harder to access than single flowers, but its anthers are still big pollen producers. Right: Dahlias are composite flowers that are popular with bees.

Plants requiring animal pollinators have to create big colourful flowers to attract insects. Insects are rewarded for the pollination service with nectar treats and nutritious pollen.

SMALLER PLANTS

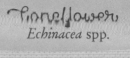
Coneflower
Echinacea spp.

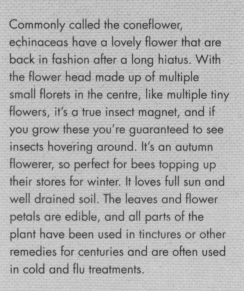

This list is by no means exhaustive. Many other plants will help your local insects by flowering and producing pollen and nectar. The goal is to select lots of different plants from the list and group them all together, with different colours, flower shapes and blooming calendars side by side, to entice different insects to feed.

Sweet alyssum
Lobularia maritima

This self-seeding annual produces white or purple flowers in a ground-cover style, which makes it a favourite stop-off for bees and butterflies. A common variety is 'Carpet of Snow', so called for the massive number of flowers it produces when it's in full bloom. Alyssums love the sun.

Sun: Full to part shade * **Soil:** Well drained * **Planting:** Pots, containers, garden * **Size:** Depending on variety, for example, 'Carpet of Snow' produces small plants less than 10 cm * **Flowers:** Summer–winter

Commonly called the coneflower, echinaceas have a lovely flower that are back in fashion after a long hiatus. With the flower head made up of multiple small florets in the centre, like multiple tiny flowers, it's a true insect magnet, and if you grow these you're guaranteed to see insects hovering around. It's an autumn flowerer, so perfect for bees topping up their stores for winter. It loves full sun and well drained soil. The leaves and flower petals are edible, and all parts of the plant have been used in tinctures or other remedies for centuries and are often used in cold and flu treatments.

Sun: Full * **Soil:** Well drained * **Planting:** * **Size:** 60–100 cm * **Flowers:** Autumn

Cornflower
Centaurea cyanus

Introduced from Europe, these annuals flower in the summer and spring and have vivid blue, thistle-like flowers that are produced for many weeks. This member of the daisy family has a number of varieties, but all have the same distinctive flower shape. They're exceptional producers of a very sweet nectar, which combined with the blue colour turns them into bee magnets.

Sun: Full to part shade * **Soil:** Well drained * **Planting:** Pots, containers, garden * **Size:** Depending on variety, up to 60 cm * **Flowers:** Late winter–summer

Cosmos
Cosmos spp.

Available as annuals and perennials, cosmos are also daisies, producing lots of flower heads. They need protection from frosts and windy conditions (so are perfect for small, closed-in areas) but also need full sun to thrive. They're very popular with all sorts of bees and other nectar- and pollen-chasing insects, and even leafcutter bees like the leaves, so it's a must for your pollinator patch.

Sun: Full to part shade * **Soil:** Well drained * **Planting:** Pots, containers, garden * **Size:** Depending on variety 45–100 cm * **Flowers:** Summer

Forget-me-not
Myosotis spp.

These spring annuals have the cutest little flowers and a distinctive fragrance that comes out at night time. Choose the blue-flowering plants for extra bee activity. A member of the borage family, it's a great container plant but a profuse spreader, so keep an eye on it.

Sun: Full to shady * **Soil:** Well drained * **Planting:** Containers, garden * **Size:** Small; can be a ground cover depending on the variety * **Flowers:** Winter–summer

Foxglove
Digitalis spp.

Foxgloves are towering plants with large spikes of long-lasting flowers that appear in spring. If you're growing them in a container you'll need a large one. They're best grown in the ground, and a healthy plant can make a great statement. With the lower lip of its tubular flowers providing a landing platform for insects, foxgloves are perfectly suited to bee pollination.

Sun: Full to part shade ✦ **Soil:** Well drained ✦ **Planting:** Large containers, garden ✦ **Size:** Tall upright flower spikes to 1.5 m tall ✦ **Flowers:** Late spring

Poppy
Papaveraceae

The red Flanders poppy is synonymous with Remembrance Day and perhaps that makes it a good one to plant for a floral display on 11 November, but is only one of a number of poppies that are easy to grow and great bee plants. The name 'poppy' refers to around 50 species of annuals and perennials originally from North America and Europe. Best grown in garden beds and in a decent display of, say, 1 square metre or more, they don't like being transplanted and most varieties are best grown from seed in place.

Sun: Full ✦ **Soil:** Well drained ✦ **Planting:** Garden ✦ **Size:** 60 cm tall ✦ **Flowers:** About 140 days after sowing in summer, winter, spring

Marigold
Tagetes spp.

Marigolds are bushy annuals that send out masses of bright blooms with big golden flower heads. There is a dwarf species (French marigold/*Tagetes patula*) available for small areas, too. Marigolds have traditionally been suggested as a good companion plant in veggie gardens, with a reputation for deterring pests through substances in their foliage. Some varieties also help control damaging nematodes, the microscopic worms that destroy the root systems of plants. Some people say marigolds are not that attractive to bees, but even if bees don't love them as much as other flowering plants, they're great companion plant in your garden to keep insect pests at bay.

Note: If you buy marigold species or varieties with open centres, insects can reach the pollen more easily.

Sun: Full ✦ **Soil:** Well drained ✦ **Planting:** Pots, containers, garden ✦ **Size:** Some varieties to 75 cm; dwarf varieties available ✦ **Flowers:** All year round in warmer parts of Australia

Rose
Rosa spp.

The magnificent rose is a great bee plant and is favoured by the leafcutter bee for nesting material. They come in all different shapes and sizes that can easily be grown in containers. Do some research before purchasing, though, as some of the hybrid varieties (popular at garden shows) have actually been bred to produce more ornate petals at the expense of pollen-producing anthers. So look for non-sterile varieties. And ignore the semi-circular holes made by the leafcutter bees. As I've said in previous chapters, it's a good thing. No really, it is! And this alone should be a good reason to grow them.

Sun: Full, sheltered from wind * **Soil:** Well drained * **Position:** Pots, containers, garden * **Size:** Everything from tall to climbers and low prostrate plants * **Flowers:** Spring–autumn

Sunflower
Helianthus annuus

Sunflowers really are statement plants. They now come in a wide assortment of colours, not just yellow, from white to rust and even several varieties of mixed shades. Sunflowers are awesome bee plants (providing nectar and pollen), they're easy to grow and kids love them.

Sun: Full to part shade * **Soil:** Deep, well drained * **Planting:** Large pots and containers, garden * **Size:** Tall plants to 2 m * **Flowers:** Summer–early autumn

Zinnia
Zinnia spp.

Another relative of the sunflower is the zinnia, named after the eighteenth-century German botanist Johann Gottfried Zinn. It's a brightly coloured, profusely flowering plant that bees adore for its huge head with abundant supplies of pollen and nectar. And it's easy to grow from seeds, too.

Sun: Full * **Soil:** Moist, well drained * **Planting:** Pots, containers, garden * **Size:** Many sizes; some varieties to 75 cm; dwarf varieties available * **Flowers:** Spring–autumn

Left page, clockwise from top: cosmos, sunflower, helenium, dahlia. This page: A garden bed of exotics.

Snapdragon
Antirrhinum spp.

My granny always had lots of snapdragons growing, and the dragonhead-shaped flowers always fascinated me. These flowers are not suited to the honeybee, but other bees such as the blue-banded bee will love them. Bumblebees, which we don't have in mainland Australia, are their natural pollinator. They really only flower in the first year, so will need to be replaced annually.

Sun: Full * **Soil:** Moist * **Position:** Pots, containers, garden * **Size:** Many sizes, but most to 30 cm * **Flowers:** Summer–autumn

Bacopa
Sutera cordata

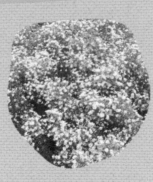

This prolific flowerer makes a great ground cover, and the hordes of compact flowers will be a massive bee magnet. Often sold as a mix of white, pink and blue flowers, it's a good one for hanging baskets, as it cascades over the sides in a lovely display.

Sun: Full to part shade * **Soil:** Moist
* **Planting:** Pots and containers are best
* **Size:** Small compact ground cover to 90 cm, depending on variety
* **Flowers:** Summer–autumn

Salvia
Salvia spp.

This flowering group, which also includes sage and mint, includes ornamental plants with more than 600 species that can vary from a low 30-cm plant to a three-metre-tall monster. The flowers can be anything from blue, purple, red, pink, white and yellow to orange. All have leaves with a very strong fragrance and release a burst of aroma when you brush past. Many varieties are winter flowerers, making them perfect to fill that nectar and pollen gap and provides winter forage.

Sun: Full to part shade * **Soil:** Moist * **Planting:** Pots, containers, garden, depending on variety * **Size:** Forms large clumps * **Flowers:** Spring–autumn

Sedum
Sedum spp.

This member of the succulent family is originally from Asia. A great variety for bees is 'Autumn Joy', which produces masses of broccoli-like flower heads that bees and birds just love. Flowering in summer to late autumn, it's another handy pre-winter larder-filler for all the beneficial insects. Being succulents, they grow well in a hot and dry spot.

Sun: Full for best flowers * **Soil:** Well drained * **Planting:** Containers, garden * **Size:** Low bush * **Flowers:** Summer–late autumn

Blue potato shrub
Solanum rantonnetii

This native of Argentina and Paraguay is a fast-growing and bushy shrub shooting up to two metres tall. It's known for its beautiful flowers that bloom purple–blue with yellow centres and attract buzz pollinators like the blue banded bee for long periods during the warm season. It's a member of the nightshade family, so don't eat the berries.

Sun: Full * Soil: Well drained * **Planting:** Pots, containers, garden * **Size:** 1.5–2 m * **Flowers:** Spring–autumn

Abelia
Abelia × *grandiflora*

This pretty ornamental shrub is named after the 1817 British consul-general in China—Dr Clarke Abel—a surgeon and naturalist. He was apparently appointed to the consulate on the suggestion of Sir Joseph Banks. This hardy plant grows just about anywhere and in any soil condition. It will grow to a height and width of about 2.5 metres and will flower for the warmer part of the year. It's a great choice for both native and European honeybees.

Sun: Full to part shade. No frost. * **Soil:** Rich, well drained * **Planting:** Pots, containers, garden * **Size:** Large bush to 1.5 m * **Flowers:** Summer–mid-autumn

Butterfly Bush
Buddleja spp.

There are about 100 species of buddleja. They're a largish bushes of 1.5 to two metres, but are available in dwarf versions as well. Some are deciduous and some even evergreen shrubs—all are renowned for their long nectar-rich white, mauve, deep purple, magenta and pink flower spikes. They're called butterfly bush because they become butterfly magnets in flower; bees and small birds will enjoy the nectar treats as well. They're hardy, quick growing, salt tolerant, and will grow in most soil types.

Sun: Full ✳ **Soil:** Well drained ✳ **Planting:** Tall pots and containers, garden ✳ **Size:** 1.5–2 m, but dwarf versions available ✳ **Flowers:** Depends on species

TREES

You don't have to think small. If you have the space then there are some great trees you can plant that will be an excellent addition to your garden, providing shade for humans, shelter for birds and forage for bees and other insects. What could be better than that?

Southern magnolia
Magnolia grandiflora

A magnificent evergreen tree originally from the United States, with huge white perfumed bee magnet blossoms, *Magnolia grandiflora* has large glossy dark green leaves with the underside a coppery velvety texture. There is also a dwarf variety called 'Little Gem' that can be grown in containers and reaches five metres tall when fully grown. Plant Grandiflora as a large statement tree, or the dwarf varieties as a hedge which can be pruned to shape.

Sun: Full to part shade. No frost or dry winds ✳ **Soil:** Well drained, drought tolerant ✳ **Planting:** Depending on the variety, in large containers or garden beds ✳ **Size:** Grandiflora up to 20 metres ✳ **Flowers:** Spring–summer

Crepe myrtle
Lagerstroemia indica

These hardy trees with a multitude of flowering colours are native to eastern Asia and grow well in most parts of Australia. (There is an Australian native species called *Lagerstroemia archeriana* with pinkish mauve flowers.) They produce amazing sprays of flowers that bees of all sorts find irresistible, especially as they flower late in summer, perfect for winter stores. Being deciduous, they're handy trees to provide extra winter sun in a corner that otherwise needs shade in summer. They're prone to suckering, though, and some older varieties also suffer from powdery mildew, so pick a newer cultivar. Crepe myrtle can be pruned back to control the size and promote flowering.

Sun: Full to part shade. No frost ✳ **Soil:** Well drained, drought tolerant ✳ **Planting:** Very large container, garden beds ✳ **Size:** up to 10 m, can be pruned to maintain a small size ✳ **Flowers:** Summer–spring

Jacaranda
Jacaranda mimosifolia

This south-central South American native is widely planted worldwide in regions with a tropical climate. It's a popular tree in Australia where it can often be seen growing in large avenues—the New South Wales town of Grafton even has a Jacaranda festival. When it flowers it produces lots and lots of purple blooms that also carpet the ground when they fall (they can be slippery and a hazard to pool filters).

The blossoms are very nectar heavy and bees can be seen taking nectar from the flowers on the tree and also on the ground, making the whole tree buzz.

Sun: Full * **Soil:** Well drained, preferably sandy * **Planting:** Garden beds * **Size:** 15 m * **Flowers:** Late spring–early summer

Chestnut
Castanea sativa

If you like roasted chestnuts then why not plant a chestnut tree and grow some of your own? They're a good nectar producer and you'll get your own nuts to roast. Chestnut trees are usually regarded as a cold-climate tree but can be grown in warmer areas if the soil is well drained. They do best if grown in pairs of different varieties to enable a good fruit yield, but beware they are large trees. There are a number of varieties available in Australia, coming from different parts of the world that suit different local climates. In Europe you can buy chestnut honey, which is often a dark strong-flavoured honey.

Sun: Depends on variety * **Soil:** Well drained, preferably sandy * **Planting:** Garden beds * **Size:** 20 m * **Flowers:** Autumn–early winter

Magnolia flowers (below) and crepe myrtle flowers (right) are both big pollen producers.

Skep hives are traditional honeybee hives made from wicker, invented about 2000 years ago. They were the beekeeper's hive of choice in medieval times.

Chapter Ten

Warre hive

BEEHIVES
AND
BEE HOTELS

A bee hotel, I hear you say? Never fear, it's not short-term accommodation. We want the bees to, ahh, bee there year upon year. Once you've reduced your chemical use and created a gourmet buffet for your bees, it's time to give them somewhere to stay.

And don't forget to add a drink spot to make your guests really happy—like many of our species, bees need a water source to survive. A pond with lilies works, so they can land on the leaves, or a plate of water full of marbles, stones or corks. Bees can easily drown in a depth of water, so dirt banks or landing platforms will become very popular. You can get very sophisticated with this sort of thing—in our bee-watering stations on our urban apiary sites we use gravel beds with float valves so they never run dry.

Don't forget to add a drinks facility for your bee guests.

NATIVE BEEHIVES —THE NEW BLACK

There's definitely a trend going on with native bees. Increasingly, people tell me they want 'some of those stingless bees' and often they're astounded when I tell them they shouldn't expect to harvest honey from them. (By all means have a taste, but don't take more than a teaspoon or so unless you live in an area where they're able to produce an excess, such as some of the northern parts of Australia.)

So before you rush out and purchase a hive of tetragonula bees, consider your environment and whether you have a suitable place for them, and how much priority you put on harvesting your own honey.

These bees are harmless and a delightful, busy addition to your backyard, providing hours of entertainment, although it does pay to keep your

mouth closed near the hive entrance. When I installed a hive of these bees at the Wayside Chapel in Sydney last year my good friend Wendy made the mistake of opening her mouth near the hive only to have a bee land on her tongue and proceed to bite her, much to the amusement of everybody except Wendy.

The site for these bees needs to be chosen carefully in a location that's not too hot and not too cold, is sheltered from the summer sun after about 10 am and from strong winds. It's essential these bees get lots of winter sun, especially in temperate climates. Another consideration is the foraging distance—these girls are limited to about one kilometre from the hive, so make sure there's plenty of food around for them. As a comparison, European honeybees can forage up to eight kilometres from the hive.

Tetragonula bee numbers have fallen in urban areas due to a loss of habitat. Installing hives can help restore ecosystem diversity and balance.

the urban beehive

Stingless Native Bees

Do some research
before installing a
Tetragonula hive. They
only have a limited
forage range and need
suitable conditions
to thrive.

If you do get a hive, then be cautious about splitting it. This involves literally ripping the hive in two to end up with two hives. Usually it works, but it's a risky move and sometimes it fails and one half or the other dies out, which leaves you and the bees feeling unhappy. Don't feel compelled to split the hive, by the way. They will be fine if you just let them be.

HOMES FOR SOLITARY BEES

I spend a lot of time working with the European honeybee. I manage honeybee hives on rooftops, balconies and community gardens as part of my business. But you don't need a beehive to contribute to bee health and plant prosperity— you can support some of your local bees in other ways. Even if you don't want to produce your own honey or keep a hive of tetragonula bees, you can contribute by creating homes for solitary bees and other beneficial insects in your garden. Habitat is in short supply for our solitary bees, and you can help them move into your backyard by providing a place to call home.

Your aim should be to replicate these habitats in a form the bees will appreciate. Most people call these 'bee hotels', and they're popping up all over the world.

Some of them are really ornate and beautifully constructed, while others are more basic, with rudimentary styling, but they all cater for the same three basic types of habitat.

You can easily make insect habitat in your backyard; it doesn't need to be fancy or have architectural merit. The insects don't mind what it looks like, as long as it provides nesting opportunities and shelter. If you want to do your bit in providing for these insects, maybe don't mention 'bees' for fear of activating the bee fearers in your household or neighbourhood.

To encourage blue-banded bees to move in, include some bee hotel suites made of clay.

Calling it an 'insect hotel' is a much better idea. Oh, and maybe don't tell them that wasps may also move into your discrimination-free habitat for fear of a revolt. The Obamas called their insect haven at the White House a butterfly garden for just that reason.

These habitats are relatively easy to construct and as I said, don't need to be a masterwork of architecture (a simple bunch of hollow sticks can make some bees very content).

Your first step is to research what sort of insects you would like to provide for, as this greatly influences how you go about constructing your bee hotel. In Australia, our native bees vary greatly in size—some are 2.5 mm long and others are 25 mm long. As a consequence they like a variety of home styles, from mud banks to holes in trees, all of which can be accommodated in your 'Four Beesons' hotel. Let's look at what sort of habitat the most common bees need.

Timber block homes provide nesting spots for solitary bees, such as leafcutter bees. Blue-banded bees will make use of them too.

Timber block homes

Many of our bees and wasps rely on borer holes in timber. These so called 'lodger' bees will also utilise any man-made hole of the correct diameter, such as old nail and bolt holes. Many gardeners and budding land-care enthusiasts remove old and rotting wood, but in doing so they're removing the habitat that our bees need. So if you have any around try to leave it there, if you possibly can.

To provide for these bees with a custom-built home, it's best to start with some new hardwood so that you can be sure it's chemical free, or use some old hardwood if you know it's clean and not going to cause your hotel to become a mausoleum. You'll be drilling holes that are about 15 cm long, so make sure the timber is big enough to allow holes this long. A minimum of 20 cm deep by 5 cm thick would be a good start. If you can't find timber the right size you could combine some blocks.

Interestingly, the depth of the hole is very important. If the holes are too short then more male bees than female bees will be raised and this will cause an imbalance in the population in your area. This may be because the males are laid before the females and there is not enough space to finish laying the females in a short tube. Solitary bees produce a nest that has compartments in it, a bit like a very short stick of bamboo. Each compartment is stocked with food and an egg that will develop into a bee. Just how long the ones at the far end need to wait before the ones in front hatch nobody knows, but I think I would prefer to be the first one out.

You want to drill holes that vary in diameter from 3 mm to 13 mm, so a selection of drill bits is a great idea, although the most common sizes are 5 mm and 6.5 mm. Auger bits are best, as they leave a clean hole, although they do tend to grab the timber so can be tricky to work with. They're also expensive. Try your local men's shed (if you're a man), as they might be able to assist and save you the expense of purchasing the drill bits. If you are a woman, then also try the local men's shed—they'll probably love an extra project and I'm sure they will lend you the tools as well. If you have only normal drill bits to work with, then with a little care they will be fine for a small number of nesting blocks.

Start by marking the holes about 2 cm apart to help prevent holes crossing each other or bursting out the side of the timber. Hardwood is by its nature hard, so drilling will be difficult. Place the wood in a vice or clamp it to something solid.

Using a piece of coloured tape, mark your drill bit at the 15 cm mark as a depth guide. Then, using your drill bit in an electric drill at a slow speed, carefully start drilling. If you're using an auger bit, as I said, they tend to grab so hang on tight. Once you're at the required depth, put the drill into reverse and back the drill out just a little bit, then put it in forward and slowly pull the drill out. This will clean the hole as it's removed. (If you pulled it all the way out by mistake just repeat.)

Once you've drilled all the holes, you'll need to clean up the entry points, as these will most likely be a bit rough. A bit of sandpaper wrapped up into a cone shape will do this. You want to make sure there are no spikes acting as a gate that will keep the bees out. A good trick is to line the holes with tubes of paper—drinking straws will do for some of them. This gives a nice clean hole for the insects to work with and they can be removed once the bees hatch and replaced with new ones. This helps keep the mites, fungal infections and other parasites under control, so your bees stay healthy, too.

You can sand the outside of the block if you wish, but don't paint it or otherwise treat the timber—it's best left to weather. If you really must, wax it with 100 per cent bees wax.

You'll want to provide a roof for your block. I often use old olive oil tins cut into a strip, but if you aren't keen on that look, then perhaps a short piece of corrugated iron bent over the block will do the trick. You'll want to position the block in a sunny spot where it's not too hot and not damp, and after time you'll see blocked holes, indicating some guests have moved in.

If there are no native bees in your area it may never get occupied, but if enough people provide this habitat then eventually the bees will make their way to you—a good reason to encourage your neighbours to give it a go, too.

Hollow reed bee homes

Quite a few of our native bees like to occupy the hollow ends of pithy plants. The reed bees (*Exoneura* spp.), which are the most common of the solitary bees, often occupy the dried ends of lantana, which I talked about in my discussion of bush regeneration issues. Unfortunately many enthusiastic people are also prone to lantana eradication frenzies, so these bees often meet an untimely demise.

Reed bees vary in length from five to eight mm and are semi-social, which means you'll find a number of females living in a single nest consisting of a hollow with no divisions. Unlike most other bees they raise all their young together.

A young bee emerging from its reed hatching chamber.

Before you rip out that lantana and burn it, try chopping off all the dried stems to 20–30 cm long, bundle them together and hang them up under a tree.

So before you rip out that lantana and burn it, try chopping off all the dried stems to 20–30 cm long, bundle them together and hang them up under a tree. The itinerant bees will hatch and if you leave it there more bees may even move in over summer.

To make more reed bee nests, or if you don't have lantana, just look for pithy-stemmed plants. All sorts of things work, such as grape vines, hydrangea and bamboo ... I've even used basil plant stems. Cut them green into lengths of 20–30 cm and varying diameters. Next, take a one-litre soft drink bottle and cut the ends off to form a tube. Now pass some string or wire through the tube to act as a hanger, bundle your sticks into fist-sized groups and pack them into the tube so they're a really tight fit. (The stems shrink when drying.) All you need to do now is hang them up in a tree or bush so they get a bit of weather protection from the leaves above, and wait for the guests to check in.

Mud-bank homes

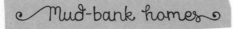

Blue-banded bees like nesting in soft earth banks and sometimes even the lime mortar in brickwork. I've had plenty of phone calls about blue-banded bees from home owners concerned their house was being permanently damaged, but in reality they don't do much harm. Vicky, my business partner in The Urban Beehive, has even attended a call out to a bee swarm only to find it was a very busy nest of blue-banded bees. These bees like to nest in groups called aggregations, and it's thought this is to evade predators because there is, literally, safety in numbers.

It's relatively easy to make habitat for these bees, but as they like living in numbers it's not always easy to get them to move in. You need a soft to firm mixture of clay and sandy soil to make the habitat, which is not always easy to find, but with a bit of experimenting you can make a mix that will hold its shape and also be soft enough to allow the bees to burrow in. Try combining three parts sand to one part clay. Maybe experiment with this mixture and make some harder and some softer mixes—the bees will move into the one they like. You'll need forms to hold the mix, and these can be sections of plastic or clay pipe or even besser blocks.

Construction is simple. You just mix up the filling, place the mould on its side and ram the mix in until it's full. To help them get going, poke a starter hole 1 cm or so deep with a pencil-sized stick. You'll need about 1 square metre of this habitat in order to give enough space for the bees to group together to make the aggregation. Just place the bee hotels all together in a sunny spot in the yard and wait a season or two. You can make smaller versions of these habitats packed into tubes—once again, old soft drink bottles work, but the tubes can be made from anything.

Grouping it all together — The Four Beesons

You can combine all of these methods together in really creative ways, making huge walls of bee habitat just waiting to be occupied. It's often easier if you make some sort of rectangular box first and then mix up the methods to make a large patch of habitat. It helps if the habitat is waterproof, so having a roof that overhangs the front is a great idea. And you can install a back piece that keeps the dampness out.

You can also construct the habitat as a wall and have it double-sided—in this way you can let

The Four Beesons—
a luxury bee hotel
offering a variety of
nesting options for the
most discerning
of guests.

Include a range of timber blocks in various shapes and sizes, with holes of various proportions, to encourage all sorts of nesting insects.

Clockwise from left: An untreated tree stump with a protective roof (pictured behind the banksia) is a simple way to provide nesting sites—drill some holes to encourage guests; a tetragonula beehive in a courtyard garden; close up of the hive entrance.

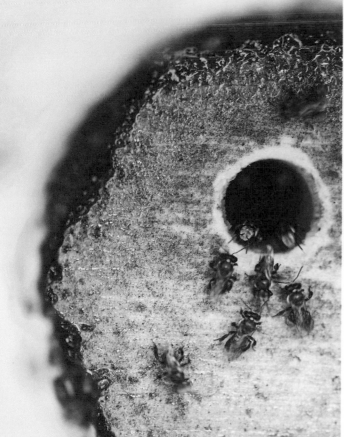

Leave strategic tree stumps around with borer holes in place and see who moves in.

creepers or vines partially cover one side for those insects that like a bit of privacy.

With these sorts of habitat, different bees and insects will often live quite happily together with no apparent territorial fights, but don't place social native beehives too close together, because they often fight to the death. About a 25-metre gap is a good rule with those hives.

You can get really creative with what you use and how you mix it up—just place the finished habitat where it gets plenty of winter sun and is sheltered from the hot summer sun. Bees and other insects use visual cues to identify their homes, so don't make it too uniform—mix up the shapes and sizes so they don't knock on the wrong door when they come home.

Just make sure it's all very secure so that birds don't steal some of the reeds and so that no reeds can fall out. Fine wire mesh works to hold it all in place, and as long as the holes in the mesh are about 15 mm, all the insects can fit through. Bird mesh is usually ideal.

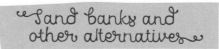

Sand banks and other alternatives

I hope I've made it abundantly clear by now that a lot of our backyards and remnant bush already contain heaps of habitat that we are clearing out in an attempt to make areas look 'human pretty' rather than 'insect pretty'. So start thinking like a bee and leave some of that rotting vegetation alone or replace it with some man-made substitutes. Your garden will be buzzing with new life before you know it. You can even make some sand banks or steps using the mixtures above for bees to try, or leave strategic tree stumps with borer holes in place around and see who moves in.

Wasps might be the unpopular cousins of bees, but they're still good pollinators. They also help keep insect pests under control, which make them great garden all-rounders.

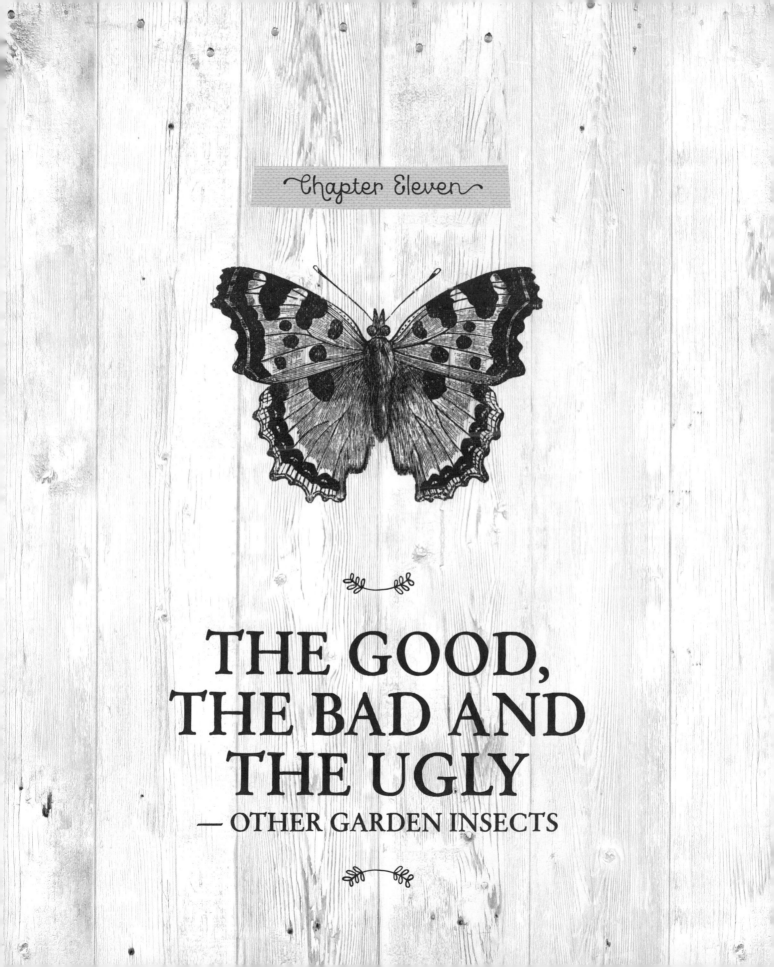

Chapter Eleven

THE GOOD, THE BAD AND THE UGLY

— OTHER GARDEN INSECTS

They're the bane of any backyard gardener: aphids on your roses, sucking the life out of them, caterpillars gnawing holes in your cabbages; wasps by the back door; snails munching your celery and stink bugs in your citrus. These slurping, chewing, stinging, biting insects are a pain when they ruin your favourite plants. So you should get rid of them all, right? Not necessarily.

If you have an unmanageable problem with a particular pest it's usually because something's out of whack in your ecosystem and needs readjusting to get your garden back into equilibrium. I've mentioned in other chapters that you shouldn't spray for pests using toxic chemicals, which I firmly believe. But, I hear you say, what's the real harm in getting rid of the pests when it gives you such a lovely pristine garden (well, at least in your human eyes anyway)?

All of our insects have roles to play. Even cockroaches. A lot are even pollinators. So before you lay down Armageddon on the bugs in your backyard you need to consider what they contribute to their ecosystem.

Many of the insects we perceive as pests in our gardens may actually be doing an important job. By keeping a pest 'under control' and thereby removing a species, we throw the garden out of balance, which sometimes allows another pest to take hold. For example, spraying aphids with a pesticide that's long-lasting will kill the bees that

Noisy miners and other birds are devoted pest controllers.

are pollinating your plants. Meanwhile, lacewings are an excellent natural predator of aphids that could take care of the problem naturally. Personally, I love seeing the noisy miner birds land in my garden every day looking for an insect snack, and will put up with caterpillars eating some of my brussels sprouts if it means I get more bird visitors.

So next time you see a bug in the garden, don't look for a way to kill it before you know what it is. Some of the ugliest-looking bugs are doing great things for your garden and really should be celebrated rather than sprayed or squashed. Your garden is a system of good and bad insects, not just bad ones, and you need all of them to be in balance to keep the garden healthy and your plants thriving.

Many of the insects we perceive as pests in our gardens may actually be doing an important job. By keeping a pest 'under control' and thereby removing a species, we throw the garden out of balance, which sometimes allows another pest to take hold.

This caterpillar of the orchard swallowtail butterfly is camouflaged to look like a bird dropping, and this helps protect it from predators.

THE GOOD GUYS

Beneficial insects are usually self-regulating, which means as predators they intentionally increase their populations when their prey (often a pest) has reached sufficient numbers to support the population, and naturally reduce or even disappear completely when their prey goes. In high population mode they will often reduce a pest population without us even knowing, leaving our fruit and veggies for us to enjoy while they've been enjoying a pest feast of their own. It's a delicate balance that can be knocked into a death spiral by an over-enthusiastic application of one chemical or another.

There are two categories of beneficial insects: predators and parasites.

Predators: Those science fiction movies with their scary, blood-thirsty monsters have nothing on the insect world. Insect predators feed on one or many different sorts of natural prey (including other insects). These bugs will eat the adults and young of their chosen targets, and sometimes supplement their diet with things like pollen and nectar. They're voracious attackers and will eat until their prey population is completely gone. They also regulate their numbers as described above. For predators to be effective you need a critical mass of their food species, otherwise they'll have insufficient numbers to make a difference.

Parasites: In the insect world parasites are usually tiny wasps or flies that invade other insects. They're more correctly called 'parasitoids' because they usually kill the insect they invade. They operate by depositing an egg into the pest, usually at an early critical life stage (a bit like the alien mother in the movie franchise *Alien*). The larva that hatches ultimately consumes and kills the host during its development, and sometimes many larvae inhabit a single individual host. Parasites are expert at finding their prey and will be effective even when the prey density is low, though they tend to be specialists and favour a single prey, which can limit their usefulness. They're pretty much invisible pest controllers, because they spend much of their development hidden away inside their host and even when they hatch are usually very small and easily missed.

Some plants are great at enticing predators into your garden, and they come from three main groups: the daisy, carrot and mint families. All of these are nectar and pollen rich, and the adult insects often rely on these sources for food, with the developing insects often the ones preying on your pests. Try planting coriander, parsley, dill, fennel, any mint variety or any daisy like aster or goldenrod, to encourage more predators into your garden.

If you don't have enough plants enticing beneficial insects in and you have plant problems, you can always buy bugs by mail order. The critters often arrive in a takeaway food container ready to be released into your garden. Follow the instructions if you do buy insects, as sometimes they can eat each other if not released in time.

Now we'll take a look at the insects you're likely to see in your backyard and talk about what they do in their predator–parasite roles.

Lacewings
Chrysopidae

These small, green insects with delicate lace wings are rapacious predators during their larval stage. They're known to eat aphids, two-spotted mite, greenhouse whitefly, scale, mealybugs, moth eggs and small caterpillars.

The females generally lay their eggs in a site that has a large number of aphids. Once the larvae hatch, off they go, swaying their heads from side to side until they touch something, which they then bite. Sometimes they're called aphid lions because of this aggressive behaviour.

The lacewing larvae develop for two to three weeks and are huge consumers of any soft-bodied insect. In the insect world they're the equivalent of Vlad the Impaler, because of their curious habit of impaling their dead on spines on their back and using them for camouflage.

Once they become adults they live on honeydew and nectar (honeydew is what aphids excrete as waste and is sometimes even collected by bees). Once the female lacewing larvae develop into adults they live for a further three to four weeks and during that time may lay hundreds of eggs each. When the new brood hatches, that's a fair number of aphid lions prowling your garden looking for tasty snacks. They can even bite humans, although I think a human would win that particular contest.

You can attract lacewings to your garden by growing specific plants and allowing some weeds their space—yes, weeds have a purpose. Try plants of the daisy family, such as dandelion (it's technically a weed so check your local regulations) or the carrot family, such as dill, angelica or flannel flowers; sunflowers; cosmos and coreopsis (considered a bush invader so be careful with this one), for starters.

You can also purchase lacewing as eggs or larvae from mail-order suppliers. The hatching lacewing larvae will eat anything, including each other, so don't wait long to place them in your garden or you may end up with one very fat grub and not much else when you open the container.

Praying mantis
Mantidae

These are among my favourite garden insects. I remember housing a few in my bug catcher as a kid. Once you see one of these you immediately understand how they got their name, they hold their front legs out in front of them like they're in prayer. They're remarkably good at camouflage and with over 100 species can resemble all kinds of vegetation—sticks, leaves and twigs. They're awesome predators, ambushing unsuspecting insects from a concealed position amongst the vegetation where they wait, front legs outstretched, completely immobile.

While they're cannibalistic, it's not true that the female eats the male after mating.

I almost never see them on my plants because they're so well camouflaged, but I do regularly see them attracted to my lights at night time.

Try planting coriander, parsley, dill, fennel, any mint variety or any daisy like aster or goldenrod, to encourage more predators into your garden.

Rove beetles
Staphylinidae

Often when disturbing the soil you'll come across a black or dark-brown beetle about 4 mm long that curves its back like a scorpion in a menacing way. It's all show and it's a rove beetle.

These little beauties live in the upper-most layer of the soil and feast on tiny soil insects like fungus gnats, which cause all sorts of damage as they tunnel into seedlings. The adult rove beetle lives on eggs and eats 50 or so fly, beetle or thrip eggs a day—a champion pest remover.

Mealybug ladybirds
Cryptolaemus montrouzieri

A great soft scale control insect is *Cryptolaemus montrouzieri*. It's an Australian native ladybird beetle that's very good at mealybug control, hence its common name. A natural predator, it's so efficient that it's earned a worldwide reputation for the job it does on mealybugs, and we export it all around the globe.

When fully grown, the insects are about 4 mm long, with an orange head and black wing covers. The ladybird larvae are hard to distinguish from mealybugs themselves, growing to about 13 mm and covered in a similar coating of waxy fur—so look twice before you squash what you think are mealybugs.

The adult beetles lay their eggs alongside mealybug eggs—a handy feast to get started on when they hatch. The female adult ladybird can lay up to 500 eggs in her life span of four to seven weeks.

Other ladybirds
Coccinellidae

Most kids can point out common ladybirds—they're the brightly coloured bugs that have been immortalised in chocolate and a stack of children's books. Few, however, would recognise the grubs that turn into ladybirds. These small, spiky, striped ugly ducklings of the insect world are what you really want to attract in your garden, as they have a huge appetite for aphids, scale insects and mites, consuming thousands a day by the time they reach maturity. That's the good news. The bad news is that some of them don't eat bugs … they're vegetarian and eat plant leaves!

There's one good vegetarian ladybird species, the fungus-eating ladybird (*Illeis galbula*) that eats mildew fungus, which is a common garden problem—you can distinguish it from other ladybirds because it's bright yellow with black markings.

The bad vegetarian (pictured) is the 28–spotted ladybird (*Henosepilachna vigintioctopunctata*), which loves the leaves of your beans, potato and cabbage plants. This pesky leaf-eating ladybird is a light orange colour and grows to about 1 cm long, and if you're not sure whether you've got them, count the spots. If you see these in your garden it's time to pick them all off and dunk them in a bucket of soapy water. One way to minimise these pests is to control the blackberry nightshade weed (*Solanum nigrum*), which is a favourite haunt and breeding ground.

Hoverfly
Syrphidae

Often confused with wasps or bees, these insects appear to be pinned in place, hence their name. You'll see them dart in, then stop in mid-air, then dart out again. The adult hoverfly relies on nectar and pollen, and is a pollinator, but it's the larvae that's most useful, especially to the aphid-plagued gardener, as the females lay their eggs among aphid colonies where the emerging larvae have a veritable aphid feast awaiting them.

Paper wasps have a tendency to build nests in inconvenient spots. This one is located outside my lounge room window. It flaps in the breeze a bit, but the wasps don't seem to mind.

Paper wasps
Polistes humilis

Wasps have a bad name in Australia. It's partly because they sting and partly because of the European paper wasp (*Vespula germanica*) species, which lives in large social colonies and can wreak havoc on a picnic or barbecue—it's the stuff of movies and 'Road Runner' cartoons. But they, like all the insect heroes of my book, serve their purpose.

The native paper wasp (not to be confused with the European wasp) often makes a small nest in urban areas under the cover of a doorway or windowsill. The nest is shaped like an upside-down cone made of wood fibre and saliva that, when dry, resembles paper and looks a bit like grey honeycomb.

I often get calls from people alarmed that bees have moved into their doorway, and asking me to remove them. It seems that if it has a honeycomb-like structure it must be a bee, so I always ask for a photo before I head out on a bee rescue mission and discover a wasp nest.

While the adult wasps eat nectar, they do feed caterpillars to their young and you can sometimes see them stuffing a caterpillar into the nest if you look closely.

They can sting and, like all wasps, don't have a barbed stinger, which means they can sting repeatedly—so don't get too close. They usually sting only when you threaten their nest, so I like to leave them alone.

Aphytis wasps
Aphytis melinus

Aphytis are tiny yellow wasps about 1.2 mm long. The adults live for a couple of weeks and each female wasp will lay 100 or so eggs during her short life.

The wasp eggs are laid under the cover of scale insects, making them an ideal predator for a range of scale, such as red scale, oriental scale and oleander scale—in fact, they can control most armoured scale.

Scale is an interesting problem caused by a small insect that builds a waxy covering (called scale) that it lives in and uses as protection. These insects also produce honeydew, which ants like to eat, and ants will effectively farm scale insects by moving them around the plant to get the best yield.

Remarkably, as the scale is impenetrable, the female wasp waits until the female scale insect is laying her eggs and has to lift up the edge of the scale, giving the wasp a chink in her armour. The wasp lays her eggs within the scale insect home, and the developing wasps feed on the scale insects as they develop. Inside their new scale insect home the wasps hatch after three weeks to begin the cycle again. Not so great for the host scale insect— it dies as the wasp larvae develop.

Whitefly parasite
Encarsia formosa

This is another tiny wasp (about 5 mm long) that feeds on the greenhouse whitefly (a common sap-sucking insect farmed by ants for its honeydew). The female wasp lays many hundreds of eggs a day inside the developing whitefly nymphs. The wasps hatch inside the fly under the protective scale, consuming the developing fly. A tell-tale sign is that the scale changes colour to black when the wasps have fed.

A number of these predator wasps are great for controlling scale type insects and aphids—it's a bug-eat-bug world out there.

Orchard swallowtail butterfly
Papilio aegeus

The orchard butterfly is one ugly caterpillar (see page 173), but a gorgeous butterfly. The grub is camouflaged to resemble a lump of bird dropping and has bright purple antennae that suddenly erupt if it's bothered. I discover these grubs on my citrus trees every year and find they do little damage. I let them be for the delight of seeing the butterfly emerge, and for other butterflies to lay eggs and start the process all over again. I was lucky enough to catch this one in the act of laying an egg.

THE BAD GUYS

These are the bugs that strike fear into the hearts of gardeners and flower fanciers across the world, and have them reaching for the spray before these pesky bugs can desecrate their favourite potted plant or bush. While it's true that these bugs can wreak havoc in the garden, there's usually a safe way to keep them under control with a regular garden pest patrol (see pest control recipes on pages 200–1).

And if this all sounds like hard work, remember that you've got a whole pest task force out there in the garden already—birds, spiders and predator insects—working around the clock for free.

Loopers and caterpillars

Cabbage loopers (*Trichoplusia ni*) are green caterpillars with several white stripes down their backs. They get their name from the way they arch their backs into a series of loops as they crawl, like cartoon caterpillars. Other types of caterpillars can also be green, or brown, and furry or smooth. Some caterpillars grow into magnificent butterflies so, as I've said a few times, make sure you have a problem, not just a couple of visitors doing minimal damage before you try to eradicate them.

The best method of control is the same with any pest—they're voracious feeders and can do lots of damage—hand-pick them and squash them or drop them in a bucket of soapy water. (Don't handle the woolly brown ones as they have irritant hairs.) You can also use *Bacillus thuringiensis*, or BT as its commonly called, a bacterium that's toxic to caterpillars but safe for humans. You can buy it in powder form from your garden shop or hardware store, then mix it with water and spray it onto the foliage the caterpillars eat. Once they consume it they die of starvation … quite a grim way to go, so I think squashing is more humane.

BT is considered safe for organic gardens and will not harm birds or other predators of the caterpillar. If you have cabbage moth caterpillars (*Pieris rapae*) you'll recognise the white moths flitting around your garden and the green caterpillars feasting on your leafy greens. You can get rid of them before they're even laid; just scatter eggshells around your plants and remove and destroy the shells, and their contents weekly. What happens is the cabbage moth apparently confuses the eggshell with a flower and lays its eggs on the shells so you can easily remove them before they hatch. Now, if only we could confuse other insects so easily …

Birds love a caterpillar meal, so you can always leave them for your fine-feathered friends. There are also a number of predators that eat caterpillar eggs, such as ladybird larvae, lacewings and even paper wasps, which take whole caterpillars and feed them to their young.

Citrus Leaf Miner
Phyllocnistis citrella

The citrus leaf miner is the larva of an introduced moth. The female moth lays her eggs on a leaf and the larvae quickly burrow between the upper and lower leaf surface of young leaves, which then become twisted or have visible white tracks under the leaf surface. Because they're damaging the leaf surface, a heavy infestation can significantly slow the growth of young trees. Larger trees can survive, with the problem becoming cosmetic only. Horticultural oil stops the moth laying her eggs, as she prefers clean leaves.

~ Aphids ~
Aphididea

Aphids are small, soft-bodied insects, yellow, green, black or red, that can be smooth or woolly. They use their long slender mouthparts to pierce stems, leaves, and other tender plants and suck out fluids, in the process dehydrating the stems and leaves, causing them to wilt, curl and go yellow. Almost all aphids have cornicles, which are a pair of tube-like structures pointing out to the rear. These are unique to the family and are a very easy way to identify them. Almost every plant has one or more aphid species that occasionally feed on it, but when the numbers grow, the plant is in trouble. And they can grow very quickly—a single aphid can produce 80 offspring in a week.

Ants don't help. They frequently farm aphids, as well as scale and mealybugs, for their sweet honeydew secretion. Like a dairy farmer and their cows, the ants often pick up their aphids and move them around when an area dries up, thus ensuring there's always a good supply of bugs. (Bees collect honeydew, too, and it's present in a lot of honey.) The presence of honeydew can sometimes indicate aphids when you haven't noticed them; and you'll often only notice honeydew is there when it gets a black sooty mould growing on it, caused by certain environmental conditions.

The easiest way to control aphids is by squashing them between your fingers, one by one. They can also be hosed off using a strong hose, but you need to do this regularly because of their potential to multiply quickly.

Before you resort to hosing or spraying, look for their predators, as biological control is very effective. The predators usually don't appear until the aphids build up in numbers, as they need the aphids as hosts or food. So look for ladybirds and larvae, lacewing larvae, soldier beetles and hoverfly larvae. And if you see aphids that have turned crusty and brown, then they have probably been affected by a parasitic wasp.

Of course, biological control is only going to work if you don't spray broad-spectrum pesticides, as the natural enemies will be killed by these sprays as well as the aphids. (See non-toxic recipes on page 200.)

If you want to try some companion planting to reduce your aphid infestation, try onions, garlic or nasturtiums, none of which will make your rose garden look grand, but at least the aphids will be kept at bay.

~ Citrus gall wasp ~
Bruchophagus fellis

If you have swollen lumps on your citrus trees, these are the egg clusters of the citrus gall wasp. The eggs are actually laid in summer and early autumn but don't hatch until winter—the lumps growing as the eggs hatch and the larvae develop. The best method of control is to remove the infected section of the branch, but don't place these in your compost. Instead, bag them up in black garbage bags and place them in the sun, or freeze them for 48 hours.

A new infestation can come from your neighbours, so it's best to get them all on board and do a neighbourhood wasp drive. Who knows, you may make some friends in the process.

Snails and slugs
Gastrpoda

These molluscs are part of
the gastropod (stomach foot)
class and prefer a damp,
vegetation-rich environment.
They like their leaves close
to the ground, and their
soil over-watered, which
is a good reason to water
your soil with a drip system
rather than spraying the
plants. You can tell when a snail's been at work versus a
slug, because they tend to consume whole young shoots
and whole plants, whereas a slug will only chew holes.
Either way, they're slimy and icky to handle. It may be
tempting to sauté up a few of your garden beauties
French style with some garlic butter, but actually—don't.
Common garden snails can harbour a few different, nasty
parasites such as lungworm, and a nematode that causes
meningitis so shouldn't be eaten. (Food-grade snails are
purged before cooking.)

Once again, fingers are the go for dispatching snails
and if you have chooks they'll love the extra food. (Snail-
flavoured scrambled eggs anybody?) Get your torch out
and go snail picking at dusk or early morning. Don't use
commercial snail baits, as most are toxic to pets, could
leach into the environment or be hazardous to wildlife.
Even iron phosphate, which breaks down into a useful
fertiliser, has now been found to be toxic to earthworms,
so don't go there. An alternative is stale beer. Yes, it's
alarming to think that there could be such a thing, but
it's very attractive to slugs and snails. You can purchase
a special dish that's sunk into the ground for the waste
brew and has a lid to keep the rain out. Or just use a
deep dish. Another alternative is crushed eggshells, grit
or sawdust. All of these stick to the snail's foot and they
don't like it so won't cross it, thereby forming a protective
barrier around your plants. If you don't mind your
chickens having a dig in your garden they also do an
excellent job of snail control.

Stink bugs
Pentatomidae

These flat-bodied, bronze, orange or green bugs are
known as stinkbugs because of the vile odour of their
secretion, designed to repel predators. They release
it from glands in their thorax when disturbed and can
squirt a fair distance. It's very caustic and can burn
foliage and eyeballs.

Stink bugs can do a lot of damage to citrus trees
as they suck sap from new shoots and flower stems,
causing shoots to wither, leaves to drop and fruit to fall
off. They're often seen in late winter or early spring. To
rid yourself of stink bugs, you can try doing what I did
recently to my lemon and lime trees—get out the vacuum
cleaner and just suck 'em up. That keeps you at arm's
length from the bugs and from the spines of whatever
tree they're sucking the life out of.

On really hot days stink bugs will migrate down the
trunk of the tree and are easier to get to. I use a wet
and dry vacuum cleaner with a little soapy water in the
bottom to drown them in. (It also helps to get the stink
bug juice off the sides of the vac.)

White Owl Grubs
Scarabaeidae

You often see these white, fleshy grubs in container gardens. They look like something out of an alien movie; the pale, prawny larvae forming a 'C' shape and curling up when disturbed. They're the larvae of the scarab or cockchafer beetle and they feed on the root systems of many plants. Unless there are heaps and heaps of them they will do little damage in your average garden. They're a significant pest around eucalypt trees, however, and are difficult to control in non-container gardens because the larvae burrow so far underground.

It's easier to control the adult beetle than the larvae by picking them off the plants, but if you are digging or re-potting and find significant numbers then it's best to re-pot with fresh mix or sieve the mix to remove the grubs.

Powdery mildews

This is a group of furry, white, warm-weather fungi that love a shady corner of the garden and thrive when the weather starts to warm up. Humid nights with a little breeze are perfect for spreading the spores. At particular risk of getting this ailment are pumpkin, cucumber, roses, grapes, strawberries and peas.

Mildew rarely kills the plant, but it can do serious damage to the leaves. There are many fungus species that cause it—all have a powdery look and can eventually grow to cover the entire leaf. Usually the lower leaves are affected first and in severe cases it can cover the entire plant, the leaves eventually dying off and yellowing in the process.

A couple of simple steps will reduce the occurrence of mildew. First, pick off the affected leaves and dispose of them in the bin so you don't spread the spores (compost needs to be really hot to kill it). Increase the air circulation around the plants by pruning and picking leaves. Don't water in the evening if you can avoid it; but if you must, try not to wet the leaves when you do. And finally, when planting, don't over-crowd.

You can use a milk spray (see recipe on page 200) but you need to cover the entire plant and spray in the morning so as to not add to the moisture.

Mites
Acari

Many different mites can be found on plants, some good and some bad, so don't just launch into a mite massacre without understanding who and what your foe is—you might kill off the good guys and create a real bad-guy problem.

If they're on your indoor plants, simply wipe the leaves with a soapy cloth. The next method is best used on outdoor plants for obvious reasons: blast them off the foliage and stems with a hose, and then use an old toothbrush to remove those clinging too tightly to be hosed off. A milk, horticultural oil or neem oil spray can also be used (see recipes on page 200).

Green planthopper
Siphanta hebes

These planthoppers are native to Australia and suck sap from leaves and branches, but don't really do much damage doing that; it's the honeydew they excrete, that can be a problem if you have lots of them. I get quite a few green planthoppers in my lime tree, which leads to a daily game of I-spy: noisy miner birds perch in the trees looking for them and then try to catch them before they hop out of harm's way. These bugs can hop a few metres, which they will do if disturbed.

Fruit Flies

There's nothing worse than finding a grub in your fruit. While some parts of the country are free from fruit fly, others have a real problem with it. There are many varieties in Australia, but really only two are a problem for backyard gardeners: the Queensland fruit fly (*Bactrocera tryoni*) on the east coast and the Mediterranean fruit fly (*Ceratitis capitata*, shown here) on the west coast. The first signs of fruit fly damage usually appear in early spring with weeping holes in developing fruit.

A good first step, which sounds a bit odd, is to control your crop production. Often we get overloaded with fruit and can't be bothered picking it, and all the fallen fruit can be a host for fruit fly. Consider pruning your tree to a manageable size or grow dwarf varieties or espaliered trees. If you have a tree that's infested or have lots of fallen fruit, collect it all, place it in black plastic bags exposed to the sun for a few days and then dispose of it in the rubbish. The heat of the sun cooks the bugs and kills them (you can use this method with weeds as well). A less smelly option is to freeze the fruit for a few days before disposing of it in the rubbish. Never put old fruit or vegetables in your compost, as they can provide a breeding platform for fruit fly.

There are both homemade options and commercial lures available for the adult flies, which will control the breeding population.

To build a homemade trap (this is not selective and may trap beneficial insects as well) take some two-litre plastic milk bottles or similar-sized containers and punch a few holes in the sides about halfway up the bottles using a hot nail or soldering iron. Make the holes about 10 mm in diameter. Put a cup of fruit juice and a bit of yeast in the bottom of each bottle to make a fermenting syrup, a bit like off fruit, which acts as an attractant (many insects find them irresistible), then hang them in your fruit trees. You'll need a few of these per tree to have the desired effect: the fruit flies enter and can't find the exit and drown.

At your garden shop you can buy pheromone traps which are designed to attract the male fruit fly, who enter the trap and perish. Reducing the number of males available to mate with means there are fewer fruit fly eggs and the cycle is disrupted. These traps are a good option, as they won't trap other possibly beneficial insects like the drowning traps can. Another option is exclusion bags, which are available in many sizes to suit your crop. These seal the fruit in and can also provide protection against other critters like birds. Most are also reusable.

Fruit fly is a neighbourhood problem, so if there are neglected trees in your area, a clean-up will benefit all local gardeners.

EVEN UGLY INSECTS
DO THEIR BIT

There are many more insects in your garden that I haven't mentioned and even whole other groups of critters, such as spiders. Many people, myself included, are creeped out by spiders, but they have a role to play and should be left alone to do it, not squashed under your boot at the first opportunity. Spiders help keep insect numbers under control by feasting on the ones that stumble into their webby trap.

I'm a lot more tolerant of spiders in my house than I used to be. A few years ago I decided to leave a huntsman spider that had taken up residence on my bedroom ceiling alone. The next night when she gave birth to thousands of baby spiders was a good lesson to me and I now put them outside with the broom rather than hitting them with it. Many people find these spiders threatening, but they prey on roaches and other insects (like mosquitoes), so they're great to have around the home to keep the numbers in balance.

Cockroaches may be a pest in your kitchen, but in the garden they help turn dead matter into mulch. (Slaters are also helpful at this job.) Some roaches are even pollinators, so if you see one in your garden get excited—it could be one of the 400 or so Australian native roaches, and working hard in your compost.

The fact is we know so little about many of the insects found in our backyards. Take flies, for instance. We know that fly maggots are responsible for breaking down dead animals, but did you know that flies are also pollinators and, according to research done by the University of New England, flies are actually better than bees at pollinating mangoes? Farmers, it seems, were already onto this and routinely dragged roadkill into their orchards to attract more flies—don't try this at home if you have a mango tree and you like a serene home life.

Flies, ants—all of them play a role and are part of the complex system of checks and balances in your garden.

Without spiders we'd face serious problems with all the plants we grow, as they're on the front line when it comes to pest control.

White-spotted rose beetle
Oxythyrea funesta

You would think with a name like that it's a beetle that eats roses, but roses are only part of its diet. It can also eat gum blossoms, grapevine, wheat, calendula, peony, thistle and daisy flowers, but while it can consume quite a few flowers it's not a huge risk to your plants. This little 8–14 mm-long beetle is also a pollinator as it spreads pollen while feasting on the flowers.

Not all new infrastructure is bad infrastructure for the environment. Set up a pollinator garden and you'll help to create a rich network for birds, bees and all the other vital insects.

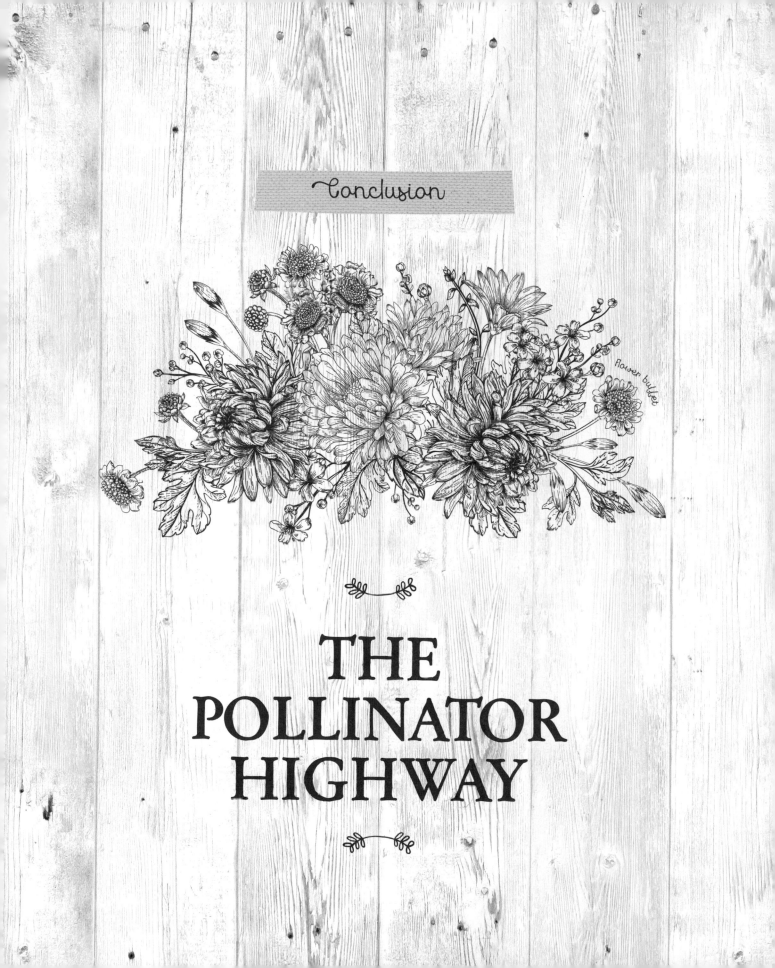

flower bullet

THE
POLLINATOR
HIGHWAY

Remember those bee goggles? By now you shouldn't need them; you're automatically seeing the world, or more specifically your garden, through a beneficial insect's eyes. But before you hand those googles on to someone else, here's a final way to think about getting things back on the right track—'the pollinator highway'.

There's a lot of new infrastructure being built these days, causing all sorts of problems to bees and wildlife, but here's some infrastructure that can actually do the planet some good.

With habitat loss occurring at an alarming rate, we can assist pollinators by planting insect-friendly gardens and providing nesting sites that will encourage the insects to repopulate areas and even spread by hopping from garden to garden—a bit like a highway. In Oslo, Norway, the first pollinator highway was planted in 2015 and is expanding rapidly. ByBi, a non-government environmental group supporting urban bees, is leading the project, and uses a website to tell people where more forage is needed. Civic-minded citizens in the area respond with a pot plant here and a garden bed there, all designed to maximise the amount of forage the local bees have available, making it easier for a bee to move from point to point. Bees can only carry so much food inside them as energy as they don't have large fat reserves like we do, so setting up lots of pollinator feeding stations means the bees can cover a larger area, taking advantage of the plentiful food along the way. It's a bit like setting up re-charging stations for an electric car.

These pollinator highways are now being established across Europe and in the United States, and now that you've read this book perhaps we will start to see some popping up in Australia as well.

The key to an effective pollinator highway is to combine efforts and talk to your neighbours, your local government, anybody who will listen, and get them on board—if it's good for the planet it's good for us. With the insects will come birds and other bad-insect predators, and before you know it your little garden oasis will be teeming with wildlife of all sorts, just like it used to be before our love affair with grass and concrete began.

That childhood you can remember that buzzed with insects is what we need to re-establish to ensure our longevity. Perhaps the bug catcher might become a favourite toy again when there are bugs to catch; and kids will start digging in the dirt as an outdoor alternative to playing Minecraft on an iPad.

Some people might be concerned that by increasing bee activity there's increased risk of being bitten or stung, but despite what your parents might have told you, creepy crawlies are not out to get you and they have as much right to be here as we do. Foraging bees are not interested in humans, they're just looking for the next flower and, like many of our insects, the sting or bite is a last-ditch effort to be left alone so they can continue doing their business.

So it's time to rip out that turf—or at least part of it—and get planting. Even a little garden or a couple of pots of flowers will have an effect. Lots of little gardens are the equivalent of one large one, and all those birds, bees and other insects will love you for it.

Let's go back to that idea from right at the beginning of the book — when you stand outside, imagine a backyard or balcony packed with flowering plants and, in the air, the chirp of crickets, the squawk of birds, and, of course, the buzz of bees, just as it's meant to be. If you could think like a bee, you'd know that bees are in trouble and its time to do something about it. It's time to become a 'beevangelist' like me, spreading the word about the bees and the flowers, before it's too late.

A European honeybee on the pollinator highway, stopping off at a magnolia feeding station.

Let's go back to that idea from right at the beginning of the book — when you stand outside, imagine a backyard or balcony packed with flowering plants and, in the air, the chirp of crickets, the squawk of birds, and, of course the buzz of bees, just as it's meant to be.

APPENDIX—
PEST CONTROL RECIPES

It's really important to remember that you should never store garden chemicals in food containers or bottles in case young children accidentally consume them. They should be stored in clearly labelled containers out of reach.

If you're considering using a particular spray, carefully research its contents and what it's used for. I've seen a chrysanthemum spray on a natural pest control website and, as I mentioned in chapter 5, chrysanthemums are what pyrethrum is made from, which is very toxic to all sorts of insects. Beware: 'natural' doesn't necessarily mean safe.

If you're going to use a spray, don't spray it on flowers, as bees could be affected when they forage. The best time to spray, if you must, is late afternoon or early evening, when the bees have finished foraging for the day and are safely at home.

Oil spray

Make a concentrate by combining 250 ml of vegetable oil and 60 ml of dishwashing soap, and mixing thoroughly. To use, shake the bottle thoroughly then measure 15 ml, or about a tablespoon, of the liquid and mix with one litre of water. Oil sprays work by suffocating the insect and are very effective against mites, scale and other soft-bodied, small insects.

Natural ant deterrent

Ants hate peppermint, so a spray bottle of water with 10–15 drops of peppermint essential oil, shaken well, will keep them away—they won't cross the sprayed area.

Milk spray

One part milk to 10 parts water, sprayed to coat the entire plant. This solution is good for all sorts of fungal problems, including black spot on roses, powdery mildew and so on. You may need to repeat this once a week if it rains.

Tip: If you use skim milk it won't smell as milky.

Herbicide alternatives for weeds

Use a blowtorch to burn the plant, or very hot water to literally cook it, or use steam—a household steamer can be a very effective weed-control measure.

If you have pavers with a recurring weed problem, try asking your paving shop for fine paving sand that you can sweep into the cracks, which fills them and makes it hard for weeds to get a hold.

Some weeds have bulbs that need to be exhausted and the only way to do this is to rob them of light using weed mats or black plastic. Eventually they will use up all their stored energy and without light be unable to make any more, so they perish.

Garlic spray

Apart from its effectiveness in repelling human suitors, garlic is a very handy garden spray for deterring soft-bodied insects, including caterpillars and aphids. The reason it works so well is that it smells and it stings ... just like onion juice in the eye, garlic juice is quite unpleasant.

A simple garlic spray can be made by steeping four cloves of garlic, crushed, with one tablespoon of dishwashing liquid in one litre of boiling water and two tablespoons of cooking oil. Steep for 24 hours, strain and away you go. Discard after two weeks, when it will begin to smell worse ... if that's possible. Be warned: The spray will irritate your eyes and nose.

INDEX

A

Abel, Dr Clarke 149
abelia 149
abiotic factors 32
Africanised honeybee 76
agastache 2–3
agriculture see food crops
alfalfa seed production 71
almond fields in California 55
almond trees 124
almond tree pollination 39, 85
Alogyne huegelii 132
American foul brood 39
angelica 175
animal manure 83
ants 181, 186
ant deterrent 200
aphids 172, 175, 179, 186
aphytis wasp 181
apple trees 26, 29–32, 125
architectural plants 14, 46
artichokes 6, 33
asbestos 54
Asian honeybee 39, 75
aster 6
Austroplebeia australis 67
avocado trees 28, 124

B

backyard fruit farms and orchards 120–125
backyard vegetable gardens 54–55, 102
bacopa 148
banana trees 125
Banks, Sir Joseph 149
Banksia spp. 136
barley 102
basil 100, 108

beans 114–115, 117
bee diseases 38–39
bee flight 84–85
bee foraging times 51, 55
bee fossil 60
bee-friendly garden examples 22–23, 30–31, 56, 90–93, 194, 198–199
bee hive 63
bee homes for native bees 158–161
bee hotels 161
bee pollinators 18, 19, 34, 35–38
bee size 60
bee species 60
bee swarms 75, 76–77
bee wings 67
beekeeping history 38–39
bee-sting deaths 51
bee-watering stations 158
beneficial insects 20, 174–183
billygoat weed 49
biotic factors 32
birds 126, 172, 184
bird baths 93
Blaauw, Brett 57
black plastic 200
blackberry nightshade 178
blackcurrant 120
blood and bone 83
blue potato shrub 149
blue-banded bee 12–14, 48, 70
blue-banded bee homes 161, 162, 164
blueberry 121
blueberry pollination 57
borage 5, 109
bottlebrush 134–135
Brachyscome spp. 128
Brazil 76

BT (*Bacillus thuringiensis*) 184
Buddleja spp. 150–151
bumblebee 38–39, 74
burrowing bee 48
Bursaria spinosa 130
butterfly bush 6, 150–151
buzz pollination 38, 70
ByBi 196

C

cabbage loopers 184
cabbage moth 85
cabbage moth caterpillars 184
Callistemon spp. 134–135
Canada 18–19
cane toad 74
carder bee 74
carrots 174, 175
caterpillars 90–92, 173, 175, 184
CCD (colony collapse disorder) 39
Ceratopetalum gummiferum 130
chalkbrood 71
chemicals in environment 14–15, 51–57
    alternatives that are questionable 83–84
    fertilisers 20, 83
    herbicides 19, 48, 51, 84
    herbicide-resistance 55
    household chemical stocktake 80–83
    insecticides 80, 81, 84
    Michigan experiment 55–57
    pesticides 46
    rise of 'cides' 54–55
    surfactants 85
chestnut 153
chives 103, 113

Christmas bush 130
chrysanthemum spray 200
citrus gall wasp 186
citrus leaf miner 185
citrus trees 121, 187
clover 51, 137
cockroaches 191
colony collapse disorder 39
commercial beehives 39
commercial herbs 102
commercial pollination 84–85
commercial vs. feral pollinators 40
community vegetable gardens
   52–53
companion planting 85, 86–87,
   104, 105, 118–119, 144, 186
coneflowers 142
   prairie 6
   purple 8, 104
constancy 35–38
containers and pots 90, 94, 95, 103, 105,
   106–107
copper chrome arsenate 51
coriander 113, 174
cornflower 143
Correa spp. 129
cosmos 143, 146, 175
crepe myrtle 39, 152, 155
crop rotation 102–105
cross-pollination 28, 32
cucumber 102

D
dahlias 33, 141, 146
daisies 24, 28, 174, 175
dandelions 90
Darwin, Charles 18
DDT 14, 54
deadly nightshade family 70
decoy plants 105
Dianella spp. 137
dill 174, 175
double flowers 140
drill bits 162–163
drones 63

E
edible plants 102
eggshells 184

eucalypts 126, 192–193
European honeybee 10, 60–65
   commercial pollination by 40
   foraging range 15
   hierarchy in hive 62–63
   at magnolia feeding station 197
   pollen baskets on 42–43
European paper wasp 181
exclusion bags 190
exotic plants 140, 147
   smaller 142–151
   trees 152–155

F
fairy fan-flowers 129
fennel 116, 174
feral bees 40, 74
fertilisers for food crops 20
fertilisers for gardens 83
flannel flowers 175
flax lily 137
flies 191
flower colour 35
flower parts 26–28
flowerhead shapes 30–31, 33
flowering all year round 90
flowering plants 14
flowering seasons 88
food crops 18, 19, 38, 40, 55
food plants 26
forget-me-not 143
Four Beesons 164–169
foxglove 144
French marigold 144
frost 20
fruit, fallen 190
fruit flies 190
fruit trees 120–125
fruit tree pollination 29–32
fungal spray 200
fungus gnats 178
fuschias 33

G
garlic spray 201
genetic diversity 18
golf courses 57
grass and lawns 46
grass and lawn alternatives 137

grass trees (Xanthorrhoea spp.) 72–73
grasses 32, 137
green manure crops 102–105
green planthopper 189
green roofs 16–17, 18
green space 46
greenhouse whitefly 175, 181
Greenpeace 19
Grevillea spp. 130
gum trees 126, 192–193

H
habitat degradation 14–15
habitat for native bees 48
habitat fragmentation 15
habitat loss 14
habitat plantings, non-insect 14, 46
hairpin banksia 136
Hardenbergia violacea 128
hay fever 32
helenium 146
herbs 90, 102, 108–113
hibiscus, native 132
hollow reed bee homes 163–164
honey 63, 64–65
honey yields 46, 47
honeycomb 35, 38, 61
honeydew 175, 181, 186, 189
horticultural oil 185
house size 12
household chemical stocktake 80–83
household cleaning products 54, 80–83
housing boom 12–14
housing landscapes 15
hoverfly 179
Hymenoptera family 60

I
insects, beneficial 20, 174–183
insect habitat 14, 46, 161, 172–173
insect parasites 174
insect pollinators 141
insect predators 174
insect repellent 80
insecticides 80, 81, 84
integrated pest management 85
introduced species 74
iron phosphate 187
Isaacs, Rufus 57

J

jacaranda 153

K

killer bee 76

L

lacewings 172, 175
ladybirds 21, 178
Lagerstroemia archeriana 152
lantana 48, 163–164
large gardens 94–97
lavender 9, 108
lawnmowing 54
leafcutter bees 42, 71
leafcutter bee homes 162
legumes 102
lemon balm 110
lemon trees 121
lemon-scented tea-tree 132
Leptospermum laevigatum 72–73
Leptospermum petersonii 132
Leptospermum polygalifolium 133
lilies 26–28
lime trees 121, 189
local council nurseries 131
local council spraying 46, 50, 57
local men's shed 162–163
lodger bees 162–163
loopers 184
lupins 6

M

macadamia pollination 67
macadamia tree 132
Magnolia grandiflora 152, 154, 197
mail order bugs 174
mango pollination 67
mango trees 191
maraschino cherries 35
marigold 144
marjoram 112
masking plants 105
mealybugs 175
mealybug ladybirds 178
Melaleuca spp. 72, 132, 133
metallic carpenter bee 72–73
Mexico 18–19
Michigan experiment 55–57

microbes 20
microclimate 94
milk spray 200
milkweed 19
mint 111, 174
mites 188
mock orange 66
monarch butterfly 18–19
moth eggs 175
mud-bank homes 12, 164

N

native bee homes 158–161
    Four Beesons 164–169
    hollow reed 163–164
    mud-bank 12, 164
    sand-bank 169
    timber block 162–163
native bee species 66–67 see also under
    individual names, eg. blue-banded bee
native cockroaches 191
native daisy 128
native fuschia 129
native green planthopper 189
native hibiscus 132
native ladybirds 178
native paper wasp 181
native plants 67
    grasses 137
    smaller 128–131
    trees 132–136
native sarsaparilla 128
native violet 129
nature strips 57, 91
nectar 12
nectar scarcity 46, 47
nectar wealth 60, 132
nectaries 14, 28, 32
nematodes 102
Newton, Isaac 80
nitrogen fixers 117
noisy miners 172
nursery plants 84, 131

O

oats 102
Obama, Barack 19
oil spray 200
onion weed 44, 83–84

Operation Pollinator 57
orange trees 121
orchard swallowtail butterfly 102, 103
orchard swallowtail caterpillar 173
oregano 112
Oslo (Norway) 196

P

paper wasps 180–181
paperbark trees (*Melaleuca* spp.) 72, 132, 133
parasitoids 174
parrots 126
parsley 103, 113, 174
passionfruit 122–123
paths 97
paving sand 200
PCB (polychlorinated biphenyl) 54
peach trees 29
peony 140
peppermint oil 200
pest control
    alternatives to chemicals 78
    integrated pest management 85
    recipes for 200–201 *see also under specific*
        pests, eg. aphids
pheromone traps 190
pigface 137
plane trees 46
plant evolution 26
planting 93–94, 98–99, 140
pollen 12
pollen baskets 42–43
pollination process
    buzz pollination 38
    flower parts 26–28
    fruit tree pollination 29–32
    pollinator syndromes 32–35
    types of pollination 32
Pollinator Health Task Force 19, 57
pollinator highway 196
pollinator syndromes 32–35
poppy 144
potatoes 102
potting mix 82
powdery mildew 102, 178, 188
praying mantis 175
professional pollination 84–85
Prostanthera lasianthos 130
public parks and gardens 12, 48

pumpkins 26, 102
purple 38
pyrethrum 80–83, 200

Q
queen bees 62

R
rapeseed 39
raspberry 121
red clover 38–39
reed bee homes 163–164
remnant bushland 46–48, 50–51
rhododendron 74
road building 15
roadside verges 57, 91
rocket 116
Rosa canina 140
Rosa rugosa 140
rosemary 109
roses 140, 145
rove beetles 178
royal jelly 62
runner beans 114–115, 117

S
sage 110
salad greens 105, 116–117
salt in weed killer 83
salvia 38, 93, 138, 148
sand-bank homes 169
Scaevola aemula 129
scale 175, 181
scent 32–35
sedum 6, 149
self-incompatible plants 28
self-pollination 26–28
single flowers 140
skep hives 156
slaters 191
slugs and snails 187
small courtyard garden beds 106–107
small gardens 94
snails and slugs 187
snapdragon 148
social bees 60
soil in purchased plants 82, 84
Solanum spp. 149, 178
solitary bees 70

solitary bee homes 161–169
sonic wand 38
sonication 38
sorghum 102
southern magnolia 152, 154
species numbers
    of bees, worldwide 60
    of grevilleas 130
    of leafcutter bees 71
    of native bees 66
    of praying mantis 175
spiders 191
spraying considerations 84–85
stale beer 187
stingless bees 66–67
stink bugs 78, 187
strawberries 40, 120
strawberry flowers 27
structural plants 14, 46
sunflower family 28, 145, 146, 175
surfactants 85
sweet alyssum 142
sweet basil 108
swimming pools 97
Syngenta 57

T
Tasmania 74
tea-trees 72–73, 132, 133
Teddy bear bee 48, 66
tennis courts 94–97
tetragonula bees 67, 68–69
tetragonula hives 159, 160, 169
Thai basil 100
thinking like a bee 12–14
thyme 97, 105, 110
timber block homes 162–163
tomatoes 102, 117
treated timber 51
tree stumps 168, 169
Tristaniopsis laurina 136
two-spotted mite 175

U
ultraviolet spectrum 35, 38
United States 18–19, 39, 57, 76
uranium smelter 54
The Urban Beehive 60
urban beehives 47

V
varroa 39
varroa mite 75
Vaucluse House 60
vegetables 116–117
vegetable gardens 54–55, 102
viewing spots 93–94, 96–97
Viola hederacea 129
von Frisch, Karl 35

W
Wallis, Captain John 60
wasps 170, 180–181, 186
water features 92
water for pollinators 93
water gum 136
watering plants 187, 188
waterlilies 33, 93
Wayside Chapel 158
weeds 46, 48, 54, 55
weed control alternatives 83–84, 200
weed killer using salt 83
weed mats 200
Wentworth, D'Arcy 60
white curl grub 188
White House butterfly garden 161
White House honey 57
whitefly parasite 175, 181
white-spotted rose beetle 192–193
wildflower planting 57
wind pollination 32
worker bees 58, 62
World Wide Fund for Nature 19
worm castings 83

Z
Zinn, Johann Gottfried 145
zinnia 145
zucchini 26, 102

ACKNOWLEDGEMENTS

I would like to thank Susan for taking a couple of long sentences and turning them into paragraphs while supporting my bee follies. The team at Murdoch—Jane, Emma and Megan—for believing in bees. Cath, for putting herself in bee peril to get some great images. Jacqui, for creating the pages. My business partner, Vicky, for doing the work of two of us while I was pounding the keyboard. The forgotten heroes of our food chain: beneficial insects everywhere. And all of you who, after reading this book, will think like a bee and plant something that flowers.

Special thanks to those who provided the locations for much of the photography in the book: Julie Hulbert, from 'Perennial Hill Garden' in Mittagong, Wendy Suma at the Wayside Chapel, Dave Jay, John and Susan Hutchinson. Thanks also to John Fairley from Country Valley for the photo showing the two fields, Gene Schembri for the photos of the native 'Four Beesons' Bee Hotel, and Katy Svalbe, Landscape Architect, Amber Road Design for help with the locations.

Published in 2016 by Murdoch Books, an imprint of Allen & Unwin
Reprinted 2017

Murdoch Books Australia
83 Alexander Street
Crows Nest NSW 2065
Phone: +61 (0) 2 8425 0100
Fax: +61 (0) 2 9906 2218
murdochbooks.com.au
info@murdochbooks.com.au

Murdoch Books UK
Erico House, 6th Floor
93–99 Upper Richmond Road
Putney, London SW15 2TG
Phone: +44 (0) 20 8785 5995
murdochbooks.co.uk
info@murdochbooks.co.uk

For Corporate Orders & Custom Publishing, contact our Business Development Team
at salesenquiries@murdochbooks.com.au.

Publisher: Jane Morrow
Design Manager: Megan Pigott
Designers: Dan Peterson and Jacqui Porter, Northwood Green
Editor: Emma Hutchinson
Photographer: Cath Muscat (see page 207 for other image credits)
Production Manager: Alexandra Gonzalez

A cataloguing-in-publication entry is available from the catalogue of the National Library of Australia at nla.gov.au.

ISBN 978 1 74336 756 8 Australia
ISBN 978 1 74336 757 5 UK

A catalogue record for this book is available from the British Library.

Colour reproduction by Splitting Image Colour Studio Pty Ltd, Clayton, Victoria
Printed by Hang Tai Printing Company Limited, China